THE NEW URBAN FRONTIER

Why have so many central and inner cities in Europe, North America and Australia been so radically revamped in the past three decades, converting urban decay into new chic? Will the process continue in the twenty-first century, or is it ended? What does this mean for the people who live there? Can they do anything about it?

This book challenges conventional wisdom – which holds gentrification to be the simple outcome of new middle-class tastes and a demand for urban living – to reveal gentrification as part of a much larger shift in the political economy and culture of the late twentieth century. Documenting in gritty detail the conflicts that gentrification brings to the new urban "frontiers," the book explores the interconnections of urban policy, patterns of investment, eviction and homelessness.

The failure of liberal urban policy and the end of the 1980s financial boom have made the end-of-the-century city a darker and more dangerous place. Public policy and the private market are conspiring against minorities, working people, the poor and homeless people as never before. In the emerging revanchist city, gentrification has become part of this policy of revenge.

Neil Smith is a Professor of Geography and Acting Director of the Center for the Critical Analysis of Contemporary Culture at Rutgers University.

THE NEW URBAN FRONTIER

Gentrification and the revanchist city

NEIL SMITH

LONDON AND NEW YORK

First published 1996
by Routledge
11 New Fetter Lane, London EC4P 4EE

Simultaneously published in the USA and Canada
by Routledge
29 West 35th Street, New York, NY 10001

Typeset in Photina by Keystroke, Jacaranda Lodge, Wolverhampton
Printed in Great Britain by
Biddles Ltd, Guildford and King's Lynn

British Library Cataloguing in Publication Data
A catalogue record for this book is available from the British Library

Library of Congress Cataloguing in Publication Data
Smith, Neil.
The new urban frontier: gentrification and the revanchist city /
Neil Smith.
p.   cm.
Includes bibliographical references and index.
1. Gentrification.   2. Gentrification – Government policy.
3. Urban renewal.   4. Urban policy.   I. Title.
HT170.S55   1996
307.76—dc20   95–46015

ISBN 0–415–13254–1 (hbk)
0–415–13255–X (pbk)

For
Nancy and Ron Smith

CONTENTS

PLATES

FIGURES

TABLES

PREFACE

In his paradigmatic essay "The significance of the frontier in American history," written in 1893, Frederick Jackson Turner (1958 edn.) proposed that:

> American development has exhibited not merely advance along a single line but a return to primitive conditions on a continually advancing frontier line, and a new development for that area. American social development has been continually beginning over again on the frontier. ... In this advance the frontier is the outer edge of the wave – the meeting point between savagery and civilization. ... The wilderness has been interpenetrated by lines of civilization growing ever more numerous.

For Turner, the expansion of the frontier and the rolling back of wilderness and savagery were an attempt to make livable space out of an unruly and uncooperative nature. This involved not simply a process of spatial expansion and the progressive taming of the physical world. The development of the frontier certainly accomplished these things, but for Turner it was also the central experience which defined the uniqueness of the American national character. With each expansion of the outer edge by robust pioneers, not only were new lands added to the American estate but new blood was added to the veins of the American democratic ideal. Each new wave westward, in the conquest of nature, sent shock waves back east in the democratization of human nature.

During the latter part of the twentieth century the imagery of wilderness and frontier has been applied less to the plains, mountains and forests of the West – now handsomely civilized – and more to US cities back East. As part of the experience of postwar suburbanization, the US city came to be seen as an "urban wilderness"; it was, and for many still is, the habitat of disease and disorder, crime and corruption, drugs and danger (Warner 1972). Indeed these were the central fears expressed throughout the 1950s and 1960s by urban theorists who focused on "blight" and "decline," "social malaise" in the inner city, the "pathology" of urban life – in short, the "unheavenly city" (Banfield 1968). The city was rendered a wilderness, or worse, a "jungle" (Long 1971; Sternlieb 1971; also Castells 1976). More vividly even than in the news media or social science narratives, this became the theme of a whole genre of Hollywood "urban jungle" movies, from *King Kong* and *West Side Story* to *The Warriors* and *Fort Apache, the Bronx*. This "discourse of decline," as Robert Beauregard (1993) has put it, dominated the treatment of the city.

Antiurbanism has been a central theme in US culture. In a pattern analogous to the original experience of wilderness, the last three decades have seen a shift from fear to romanticism and a progression of urban imagery from wilderness to frontier. Cotton Mather and the Puritans of seventeenth-century New England feared the forest as an impenetrable evil, a dangerous wilderness, a primeval place. But with the continual taming of the forest and its transformation at the hands of increasingly capitalized human labor, the softer imagery of Turner's frontier became an obvious successor to Mather's forest of evil. There is an optimism and an expansive expectation associated with the "frontier" which refracts a sense of self-confident conquest. Thus in the twentieth-century US city, the imagery of urban wilderness – a desperate relinquishing of hope – was beginning, by the 1960s (widespread uprisings notwithstanding), to be replaced by a vision of the urban frontier. This transformation can be traced in part to the discourse of urban renewal (Abrams 1965), but was intensified in the 1970s and 1980s as the rehabilitation of single-family homes and tenement blocks became increasingly symbolic of a successor form of "urban renewal." In the language of gentrification, the appeal to frontier imagery has been exact: urban pioneers, urban homesteaders and urban cowboys became the new folk heroes of the urban frontier. In the 1980s, the real estate magazines even talked about "urban scouts" whose job it was to scout out the flanks of gentrifying neighborhoods, check the landscape for profitable reinvestment, and, at the same time, to report home about how friendly the natives were. Less optimistic commentators indict the emergence of a new group of "urban outlaws" in connection with inner-city drug cultures.

Just as Turner recognized the existence of Native Americans but included them as part of his savage wilderness, contemporary urban frontier imagery treats the present inner-city population as a natural element of their physical surroundings. The term "urban pioneer" is therefore as arrogant as the original notion of "pioneers" in that it suggests a city not yet socially inhabited; like Native Americans, the urban working class is seen as less than social, a part of the physical environment. Turner was explicit about this when he called the frontier "the meeting point between savagery and civilization," and although the 1970s and 1980s frontier vocabulary of gentrification is rarely as explicit, it treats the inner-city population in much the same way (Stratton 1977).

The parallels go further. For Turner, the westward geographical progress of the frontier line is associated with the forging of the "national spirit." An equally spiritual hope is expressed in the boosterism which presents gentrification as the leading edge of an urban renaissance; in the most extreme scenario, the new urban pioneers were expected to do for the flagging national spirit what the old ones did: to lead the nation into a new world where the problems of the old world are left behind. In the words of one federal publication, gentrification's appeal to history involves the "psychological need to re-experience successes of the past because of disappointments of recent years – Vietnam, Watergate, the energy crisis, pollution, inflation, high interest rates, and the like" (Advisory Council on Historic Preservation 1980). From here, as we shall see, it was a short path from a failed liberalism to the revanchist city of the 1990s. No one has yet seriously proposed that we view James Rouse – the developer responsible for such maverick downtown tourist arcades as Baltimore's Inner Harbor, South Street Seaport in New York or Boston's Fanueil Hall – as the John Wayne of gentrification, but insofar as such projects serve to anchor the gentrification of many down-towns, the proposal would be quite in keeping with the frontier discourse. In the end, and this is the important conclusion, the frontier discourse serves to rationalize and legitimate a process of conquest, whether in the eighteenth- and nineteenth-century West, or in the late-twentieth-century inner city.

Turner's effect on Western history is still monumental, and the groove he carved for a patriotic history has been difficult to escape. Yet a new generation of "revisionist" historians has begun to rewrite the history of the frontier. Patricia Nelson Limerick senses the latter-day urban reappropriation of the frontier motif in her corrective to Hollywood histories of the West:

> If Hollywood wanted to capture the emotional center of western history, its movies would be about real estate. John Wayne would have been neither a gunfighter nor a sheriff, but a surveyor, speculator or claims lawyer. The showdowns would appear in the land office or the court-room; weapons would be deeds and lawsuits, not six-guns.
>
> (Limerick 1987: 55)

Now this might seem in many ways a highly nationalist scripting of the gentrification process. In fact, of course, gentrification is a thoroughly inter-national phenomenon, having emerged widely in the cities of Canada, Australia, New Zealand and Europe, and more sporadically in Japan, South Africa and Brazil. In Prague or Sydney, or for that matter Toronto, the language of the frontier is not such an automatic ideological lubricant to gentrification as in the US, and this frontier mythology, applied to the *fin-de-siècle* city, seems a distinctly American creation. While there is no doubt that the frontier mythology is more viscerally present in the US, still the original frontier experience is not simply a US commodity. In the first place, it was as intensely real a vision of the New World for potential immigrants from Scandanavia or Sicily as it was for the Germans or Chinese already living in Kansas City or San Francisco. But second, other European-colonial outposts – the Australian or Kenyan outback, the "northwest frontier" of Canada or India and Pakistan, for example – shared different but equally powerful elixirs of frontier and class, race and geography which laid them open to parallel ideologies. And finally, the frontier motif has in any case emerged in non-US situations.

Most notably, perhaps, the frontier emerges in London as what became known as the "frontline". Following riots between police and Afro-Caribbean, South Asian, and white youths in London (and other British cities) through the 1980s, a territorial line emerged in several neighborhoods. These frontlines, such as All Saints Road in Kensington and Chelsea (Bailey 1990) or those in Notting Hill or Brixton, were simultaneously defenses against police incursions in the 1970s and at the same time strategic "beachheads" established by the police. They also quickly became antigentrification lines in the 1980s. Sir Kenneth Newman, former Metropolitan Police commissioner, launched the police dimension of this frontline strategy in the early 1980s, and explained its purpose in a lecture to the right-wing European Atlantic Group. Citing the "growth of multi-ethnic communities" which were responsible for producing a "deprived underclass," Newman anticipated "crime and disorder," and identified eleven "symbolic locations" in London, including the frontlines, where special tactics would be required. For each location, "there was a contingency plan to enable the police swiftly to occupy the area and exert control" (cited in Rose 1989).

The frontier motif has been every bit as literal in the cultural froth of everyday London life. As enthusiastically as anywhere in the US, "urban cowboy" became a cult style for some. "Yup, it's high noon all over London," says Robert Yates (1992) "and Wild West fanatics are donning their stetsons, saddling their horses, and making believe that Tower Bridge is Texas." In Copenhagen the "Wild West Bar" opens in a gentrified neighborhood where, in May 1993, six protestors were shot by police in a riot following a Danish vote for the Maastricht European Union treaty. From Sydney to Budapest, Wild West bars and other frontier symbols regularly script and adorn the gentrification of city neighborhoods. And of course the motif often sports a distinctive local moniker, as with the empire theme in London wherein the gentrifiers become "the new Raj" (M. Williams 1982) and the "Northwest Frontier" takes on an entirely new symbolic and political meaning (see also Wright 1985: 216–248). In this version, the internationalism of gentrification is more directly admitted.

As with every ideology, there is a real and partial if distorted basis for the treatment of gentrification as a new urban frontier. The frontier represents an evocative combination of economic, geographical and historical advances, and yet the social individualism pinned to this destiny is in one very important respect a myth. Turner's frontier line was extended westward less by individual pioneers, homesteaders, rugged individualists, than by banks, railways, the state and other collective sources of capital (Swierenga 1968; Limerick 1987). In this period, economic expansion was accomplished largely through geographical expansion on a continental scale.

Today the link between economic and geographical expansion remains, giving the frontier imagery its potency, but the form of that connection is very different. Economic expansion today no longer takes place purely via absolute geographical expansion but rather involves internal differentiation of already developed spaces. At the urban scale, this is the importance of gentrification vis-à-vis suburbanization. The production of space in general and gentrification in particular are examples of this kind of uneven development endemic to capitalist societies. Much like a real frontier, the gentrification frontier is advanced not so much through the actions of intrepid pioneers as through the actions of collective owners of capital. Where such urban pioneers go bravely forth, banks, real estate developers, small-scale and large-scale lenders, retail corporations, the state, have generally gone before.

In the context of so-called globalization, national and international capitals alike confront a global "frontier" of their own that subsumes the gentrification frontier. This link between different spatial scales, and the centrality of urban development to national and international expansion, was acutely clear in the enthusiastic language of supporters of urban Enterprise Zones, an idea pioneered by the Thatcher and Reagan governments in the 1980s, and a centerpiece of 1990s urban privatization strategies. As Stuart Butler (a British economist working for the extreme right-wing American think tank, the Heritage Foundation) suggests, in this diagnosis of urban malaise, the conversion of the inner city into a frontier is not an accident and the imagery is more than a convenient ideological conveyance. As in the nineteenth-century West, the construction of the new urban frontier of the *fin de siècle* is a political geographical strategy of economic reconquest:

> It may be argued that at least part of the problem facing many urban areas today lies in our failure to apply the mechanism explained by Turner (the continual local development and innovation of new ideas) to the inner city "frontier". . . . Proponents of the Enterprise Zone aim to provide a climate in which the frontier process can be brought to bear within the city itself.
>
> (Butler 1981: 3)

* * *

This book is divided into four parts. The introductory part sets the stage for the social, political and economic conflicts that are raised by gentrification. The first chapter focuses on the struggle over Tompkins Square Park in New York's Lower East Side, and highlights how one of the most intense anti-gentrification struggles of the 1980s turned the neighborhood into a new urban frontier. The second chapter offers a short history of gentrification and a survey of current debates, and makes the central argument that in the 1990s, continuing gentrification contributes to what I call the "revanchist city." Part I pieces together several theoretical strands that help explain gentrification. Where chapter 3 focuses on the housing market and the local scale, chapter 4 is explicitly global in focus and deals with wider economic arguments about uneven development. Chapter 5 considers some of the arguments that connect gentrification with the social restructuring of class and gender. Using case studies from Philadelphia, Harlem, Budapest, Amsterdam and Paris, Part II attempts to show the fluid interconnection between global shifts in the social economy and the myriad detail of local instances of gentrification. I emphasize here the role of the state and the "catch 22" character of the process for existing working class residents, as well as the different contours of gentrification in different cities and different decades. Part III attempts to turn the frontier motif on its head. By actually mapping the gentrification frontier, we can demonstrate the kernel of harsh economic geography around which the acclamatory cultural scripting of urban pioneering is built. The last chapter argues that the emerging revanchist urbanism of the *fin de millenaire* city, especially in the United States, embodies a revengeful and reactionary viciousness against various populations accused of "stealing" the city from the white upper classes. Gentrification, far from an aberration of the 1980s, is increasingly re-emerging as part of this revanchism, an effort to retake the city.

* * *

In retrospect I suppose I first saw gentrification in 1972 while working for the summer in an insurance office in Rose Street in Edinburgh. Every morning I took the 79 bus in from Dalkeith and walked half the length of Rose Street to the office. Rose Street is a back street off majestic Princes Street and long had a reputation as nightspot with some long-established traditional pubs and a lot of more dingy howffs – watering holes – and even a couple of brothels, although these were rumored to have decamped to Danube Street by the early 1970s. It was *the* place in Edinburgh for a pub crawl. My office was above a new bar called "The Galloping Major" which had none of the cheesy decor or sawdust on the floor of the old-time bars. This one was new. It served quite appetizing lunches adorned with salad, still a novelty in most Scottish pubs at the time. And I began to notice after a few days that a number of other bars had been "modernized"; there were a couple of new restaurants, too expensive for me – not that I went to restaurants much in any case. And narrow Rose Street was always clogged with construction traffic as some of the upper floors were renovated.

I didn't think much of this at the time, and only several years later in Philadelphia, by which time I had picked up a little urban theory as a geography undergraduate, did I begin to recognize what I was seeing as not only a pattern but a dramatic one. All the urban theory I knew – which wasn't much, to be sure – told me that this "gentrification" wasn't supposed to be happening. Yet here it was – in Philadelphia *and* Edinburgh. What was going on? In the remaining years of the 1970s I had many similar experiences. I heard and loved Randy Newman's song, "Burn on Big River" as a biting environmental protest, but by the time I got to Cleveland in 1977, the bar scene in the Flats by the Cuyahoga River was already beginning to attract a few yuppies, and students like myself, as well as Hell's Angels and the last workers from the docks. I thought I saw the writing on the wall. I bet an incredulous friend from Cleveland that the city would have significant gentrification within ten years, and although she never did fork over, she was forced to admit defeat long before the ten years were up.

The essays in this book involve a variety of experiences of gentrification but they are based more in the US than anywhere else. Indeed three or four of the chapters – especially the concluding arguments discussing political and cultural opposition to gentrification – are based on my experiences and research in New York City. This obviously raises questions about the applicability of the arguments in other contexts. While I accept the admonition that radically different experiences of gentrification obtain in different national, regional, urban and even neighborhood contexts, I would also hold that among these differences a braid of common threads ripples through most experiences of gentrification. A lot can be learned from the New York experience, and there is much in New York that strikes a chord elsewhere. When Lou Reed sang "Meet You in Tompkins Square" (on his album *New York*) he made the violent struggles around that park in the Lower East Side an instantly recognizable international symbol of the emerging "revanchist city" for many people.

Many of the chapters in this book represent revised and edited versions of essays I have published previously, and so my first debt lies with my co-authors. I am especially grateful to Richard Schaffer, with whom I worked on the initial Harlem research in Chapter 7, and to Laura Reid and Betsy Duncan, who coauthored a first version of Chapter 8. I should also acknowledge a National Science Foundation Grant no. SE-87-13043 which sponsored the research reported in Chapter 9.

Many people have commented on different aspects of this work and in other ways contributed to it. The following list is very partial and I apologize for those I have inevitably missed in reconstructing the history: Rosalyn Deutsche, Benno Engels, Susan Fainstein, David Harvey, Kurt Hollander, Ron Horvath, Andrea Katz, Hal Kendig, Les Kilmartin, Larry Knopp, Mickey Lauria, Sheila Moore, Damaris Rose, Chris Tolouse, Michael Sorkin, Ida Susser, Leyla Vural, Peter Williams, Sharon Zukin. Many people have eagerly introduced me to gentrification in their cities and helped to broaden my own vision: Benno Engels, Ron Horvath, Janelle Allison, Ruth Fincher, Mike Webber, Blair Badcock, Judit Timár, Viola Zentai, Zoltan Kovács, Ed Soja, Helga Leitner, Eric Sheppard, Jan van Weesep, John Pløger, Anne Haila, Alan Pred, Eric Clark, Ken and Karen Olwig, Steen Folk.

I am grateful to Mike Siegel, who drew the maps and artwork, and to Ruthie Gilmore, Marla Emory, Annie Zeidman, and especially Tamar Rothenberg, who gave excellent research support at various stages. Whatever coherence the book can claim owes a lot to them.

Several people have been especially important in my gentrification research. Roman Cybriwsky was very generous with his time, ideas and support at the earliest stage of my gentrification research, and that generosity continues in his donation of a print for this book. Briavel Holcomb has been an equally generous and supportive colleague, always passing something on that will interest me – including brown-eared copies of her confidential referee's reports on some of my earliest work. Bob Beauregard always somehow found the time to fully engage, even when he disagreed; with Bob Lake and Susan Fainstein, Bob Beauregard has been the most collegial of colleagues.

Eric Clark has been a firm critic as well as supporter; I have learned a lot from his arguments, on paper and in person, and have benefited from his generosity. Jan van Weesep invited me to Utrecht in 1990, and thereby gave me the time and space to begin thinking about gentrification in a broader context. But not before he organized a two-day conference on "European gentrification" – then promptly lent me his car on the second day to explore the Polders (where there is no gentrification) so that he could get out a

European gentrification agenda, unobstructed by my insistence on a global purview. A fair exchange is no robbery. Chris Hamnett, who arrived in Utrecht along with the first squalls of a hurricane, is a longtime friend and impish antagonist without whom discussions of gentrification would have been a lot more bromidic.

I have to make special acknowledgment of Joe Doherty. At a particularly impressionable point in my education I had the enthusiastic idea of studying the diffusion of new silage technologies in the Midwest, and without Joe's gentle and patient guidance that gentrification was something I could get my teeth into, I might well have become a pastoral geographer. In the same context, I should also acknowledge the bureaucrat in the US Department of Agriculture – I forget his name – who never answered my letter requesting data, and thereby made Joe's advice more persuasive. Joe was also the one who alerted me to Ruth Glass's role in coining the term "gentrification."

Rick Schroeder, Do Hodgson, Tim Brennan, David Harvey, Haydee Salmun, Delfina Eva Harvey, Ruthie Gilmore, Craig Gilmore, Sallie Marston are friends whose influence, support and comradeship transcend any concern with gentrification. Indeed, they remind me that there is life after gentrification, although I am not always so sure.

I have known Cindi Katz for just about as long as I have known gentrification, but only since the day that the New York City police first violently evicted homeless people from Tompkins Square Park, on the coldest day of December 1989, have Cindi and gentrification been entwined together in my life. With her I would love to see a world after gentrification, and a world after all the economic and political exploitation that makes gentrification possible: the personal tendrils of a new politics.

Finally, the journey from Dalkeith to Philadelphia in 1974 was very much a journey away from home. With this book I might be able to give something back. Dalkeith isn't quite facing gentrification, I suspect, but most people in Dalkeith will recognize the broad politics of gentrification only too well. Therefore I would like to dedicate this book to my mother and father, Nancy and Ron Smith, who ensured not just that I got the education that took me away but that it was a political education. I know they will be honored to share the dedication with people fighting gentrification everywhere.

ACKNOWLEDGMENTS

Many of the chapters in this book are revised, updated and edited versions of papers that first saw the light of day elsewhere. They appear here with the permission of those editors and publishers. Chapter 1 is an updated adaptation of two esays: "Tompkins Square: riots rents and redskins," *Portable Lower East Side* 6 (1989), edited by Kurt Hollander; and "New city, new frontier: the Lower East Side as Wild West," in Michael Sorkin (ed.) *Variations on a Theme Park: The New American City and the End of Public Space*, New York: Hill and Wang (1992). Chapter 3 is derived from "Toward a theory of gentrification: a back to the city movement by capital not people," *Journal of the American Planning Association* 45 (1979). An earlier version of Chapter 4 was originally published as "Gentrification and uneven development," *Economic Geography* 58 (1982). Chapter 5 expands upon "Of yuppies and housing: gentrification, social restructuring and the urban dream," *Environment and Planning D: Society and Space* 5 (1987). Chapter 6 originally appeared as "Gentrification and capital: theory, practice and ideology in Society Hill," *Antipode* 11, 3 (1979). An earlier version of Chapter 7 was coauthored with Richard Schaffer: "The gentrification of Harlem?", *Annals of the Association of American Geographers* 76 (1986). Chapter 9 is distilled from "From disinvestment to reinvestment: tax arrears and turning points in the East Village," *Housing Studies* 4 (1989), coauthored with Laura Reid and Betsy Duncan. And Chapter 10 develops the theme of two papers: "After Tompkins Square Park: degentrification and the revanchist city," in A. King (ed.) *Re-presenting the City: Ethnicity, Capital and Culture in the 21st Century Metropolis*, London: Macmillan (1995) and "Social justice and the new American urbanism: the revanchist city," in Eric Swyngedouw and Andrew Merrifield (eds.) *The Urbanization of Injustice*, London: Lawrence and Wishart (1996).

INTRODUCTION

"CLASS STRUGGLE ON AVENUE B"

The Lower East Side as Wild Wild West

On the evening of August 6, 1988, a riot erupted along the edges of Tompkins Square Park, a small green in New York City's Lower East Side. It raged through the night with police on one side and a diverse mix of anti-gentrification protestors, punks, housing activists, park inhabitants, artists, Saturday night revelers and Lower East Side residents on the other. The battle followed the city's attempt to enforce a 1:00 A.M. curfew in the Park on the pretext of clearing out the growing numbers of homeless people living or sleeping there, kids playing boom boxes late into the night, buyers and sellers of drugs using it for business. But many local residents and park users saw the action differently. The City was seeking to tame and domesticate the park to facilitate the already rampant gentrification on the Lower East Side. "GENTRIFICATION IS CLASS WAR!" read the largest banner at the Saturday night demonstration aimed at keeping the park open. "Class war, class war, die yuppie scum!" went the chant. "Yuppies and real estate magnates have declared war on the people of Tompkins Square Park," announced one speaker. "Whose fucking park? It's our fucking park," became the recurrent slogan. Even the habitually restrained *New York Times* echoed the theme in its August 10 headline: "Class War Erupts along Avenue B" (Wines 1988).

In fact it was a *police* riot that ignited the park on August 6, 1988. Clad in space-alien riot gear and concealing their badge numbers, the police forcibly evicted everyone from the park before midnight, then mounted repeated baton charges and "Cossacklike" rampages against demonstrators and locals along the park's edge:

> The cops seemed bizarrely out of control, levitating with some hatred I didn't understand. They'd taken a relatively small protest and fanned it out over the neighborhood, inflaming hundreds of people who'd never gone near the park to begin with. They'd called in a chopper. And they would eventually call 450 officers. . . . The policemen were radiating hysteria. One galloped up to a taxi stopped at a traffic light and screamed, "Get the fuck out of here, fuckface. . . . " [There were] cavalry charges down East Village streets, a chopper circling overhead, people out for a Sunday paper running in terror down First Avenue.
>
> (Carr 1988: 10)

Plate 1.1 New York City police retake Avenue A on the edge of Tompkins Square Park. 1988 (© Andrew Lichtenstein)

Finally, a little after 4:00 A.M. the police withdrew in "ignominious retreat," and jubilant demonstrators reentered the park, dancing, shouting and celebrating their victory. Several protestors used a police barricade to ram the glass-and-brass doors of the Christodora condominium, which borders on the park on Avenue B and which became a hated symbol of the neighborhood's gentrification (Ferguson 1988; Gevirtz 1988).[1]

In the days following the riot, the protestors quickly adopted a much more ambitious political geography of revolt. Their slogan became "Tompkins Square everywhere" as they taunted the police and celebrated their liberation of the park. Mayor Edward Koch, meanwhile, took to describing Tompkins Square Park as a "cesspool" and blamed the riot on "anarchists." Defending his police clients, the president of the Patrolmen's Benevolent Association enthusiastically elaborated: "social parasites, druggies, skinheads and communists" – an "insipid conglomeration of human misfits" – were the cause of the riot, he said. In the following days, the city's Civilian Complaint Review Board received 121 complaints of police brutality, and, largely on the evidence of a four-hour videotape made by local video artist Clayton Patterson, seventeen officers were cited for "misconduct." Six officers were eventually indicted but none was ever convicted. The police commissioner only ever conceded that a few officers may have become a little "overenthusiastic" owing to "inexperience," but he clung to the official policy of blaming the victims (Gevirtz 1988; Pitt 1989).

Prior to the riot of August 1988, more than fifty homeless people, evictees from the private and public spaces of the official housing market, had begun to use the park regularly as a place to sleep. In the months following, the number of evictees settling in the park grew, as the loosely organized antigentrification and squatters' movements began to connect with other local housing groups. And some of the evictees attracted to the newly "liberated space" of Tompkins Square Park also began to organize. But the City also slowly regrouped. City-wide park curfews (abandoned after the riot) were gradually reinstated; new regulations governing the use of Tompkins Square Park were slowly implemented; several Lower East Side buildings occupied by squatters were demolished in May 1989, and in July a police raid destroyed tents, shanties and the belongings of park residents. By now there were on average some 300 evictees in the park on any given night, at least three-quarters men, the majority African-American, many white, some Latino, Native Americans, Caribbean. On December 14, 1989, on the coldest day of the winter, the park's entire homeless population was evicted from the park, their belongings and fifty shanties hauled away into a queue of Sanitation Department garbage trucks.

It would be "irresponsible to allow the homeless to sleep outdoors" in such cold weather, explained a disingenuous parks commissioner, Henry J. Stern, who did not mention that the city shelter system had beds for only a quarter of the city's homeless people. In fact, the city's provision for the evicted ran only to a "help center" that, by one account, "proved to be little more than a dispensary for baloney sandwiches" (Weinberg 1990). Many evictees from the park were taken in by local squats, others set up encampments in the neighborhood, but quickly they filtered back to Tompkins Square. In January 1990 the administration of supposedly progressive mayor David Dinkins felt sufficiently confident of the park's eventual recapture that it announced a "reconstruction plan." In the next summer the basketball courts at the north end were dismantled and rebuilt with tighter control of access; wire fences closed off newly constructed children's playgrounds; and park regulations began to be more strictly enforced. In an effort to force evictions, City agencies also heightened their harassment of squatters who now spearheaded the anti-gentrification movement. As the next winter closed in, though, more and more of the city's evictees came back to the park and began again to construct semipermanent structures.

In May 1991, the park hosted a Memorial Day concert organized under the slogan "Housing is a human right" and, in what was becoming an annual May ritual, a further clash with park users ensued. It was now nearly three years since protestors had taken the park, and, with almost a hundred shanties, tents and other structures now in Tompkins Square, the Dinkins administration decided to move. The authorities finally closed the park at 5:00 A.M. on June 3, 1991, evicting between 200 and 300 park dwellers. Alleging that Tompkins Square had been "stolen" from the community by "the homeless," Mayor Dinkins declared: "The park is a park. It is not a place to live" (quoted in Kifner 1991). An eight-foot-high chain-link fence was erected, a posse of more than fifty uniformed and plainclothes police was delegated to guard the park permanently – its numbers augmented to several hundred in the first days and during demonstrations – and a $2.3 million reconstruction was begun almost immediately. In fact, three park entrances were kept open and heavily guarded: two provided access to the playgrounds for children only (and accompanying adults); the other, opposite the Christodora condominium, provided access to the dog run. The closure of the park, commented *Village Voice* reporter Sarah Ferguson, marked the "death knell" of an occupation that "had come to symbolize the failure of the city to cope with its homeless population" (Ferguson 1991b). No alternative housing was offered evictees from the park; people again moved into local squats, or filtered out into the city. On vacant lots to the east of the park, a series of shantytown communities were erected and they quickly took the name "Dinkinsville," linking the present mayor with the "Hoovervilles" of the Depression. Dinkinsville was less a single place than a collection of communities, with a similar impossible geography to that of Bophuthatswana. Existing collections of shanties under the Brooklyn, Manhattan and Williamsburg Bridges expanded.

As the site of the most militant antigentrification struggle in the United States (but see Mitchell 1995a), the ten acres of Tompkins Square Park quickly became a symbol of a new urbanism being etched on the urban "frontier." Largely abandoned to the working class amid postwar suburban expansion, relinquished to the poor and unemployed as reservations for racial and ethnic minorities, the terrain of the inner city is suddenly valuable again, perversely profitable. This new urbanism embodies a widespread and drastic repolarization of the city along political, economic, cultural and geographical lines since the 1970s, and is integral with larger global shifts. Systematic gentrification since the 1960s and 1970s is simultaneously a response and contributor to a series of wider global transformations: global economic expansion in the

Plate 1.2 The closing of Tompkins Square Park, June 3, 1991 (© Andrew Lichtenstein)

Plate 1.3 Tompkins Square Park fenced off, 1992 (© Andrew Lichtenstein)

1980s; the restructuring of national and urban economies in advanced capitalist countries toward services, recreation and consumption; and the emergence of a global hierarchy of world, national and regional cities (Sassen 1991). These shifts have propelled gentrification from a comparatively marginal preoccupation in a certain niche of the real estate industry to the cutting edge of urban change.

Nowhere are these forces more evident than in the Lower East Side. Even the neighborhood's different names radiate the conflicts. Referred to as *Loisaida* in local Puerto Rican Spanish, the Lower East Side name is dropped altogether by real estate agents and art world gentrifiers who, anxious to distance themselves from the historical association with the poor immigrants who dominated this community at the turn of the century, prefer "East Village" as the name for the neighborhood above Houston Street. Squeezed between the Wall Street financial district and Chinatown to the south, the Village and SoHo to the west, Gramercy Park to the north and the East River to the east (Figure 1.1), the Lower East Side feels the pressure of this political polarization more acutely than anywhere else in the city.

Highly diverse but increasingly Latino since the 1950s, the neighborhood was routinely described in the 1980s as a "new frontier" (Levin 1983). It mixes spectacular opportunity for real estate investors with an edge of daily danger on the streets. In the words of local writers, the Lower East Side is variously a "frontier where the urban fabric is wearing thin and splitting open" (Rose and Texier 1988: xi) or else "Indian country, the land of murder and cocaine" (Charyn 1985: 7). Not just supporters but antagonists have found this frontier imagery irresistible. "As the neighborhood slowly, inexorably

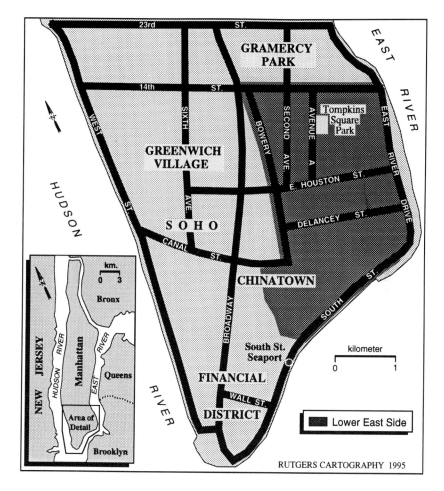

Figure 1.1 New York's Lower East Side

gentrifies," wrote one reporter, in the wake of the 1988 police riot, "the park is a holdout, the place for one last metaphorical stand" (Carr 1988: 17). Several weeks later, "Saturday Night Live" made this Custer imagery explicit in a skit cast in a frontier fort. Custer (as Mayor Koch) welcomes the belligerent warrior Chief Soaring Eagle into his office and inquires: "So how are things down on the Lower East Side?"

The social, political and economic polarization of "Indian country" is drastic and fast becoming more so. Apartment rents soared throughout the 1980s and with them the numbers of homeless; record levels of luxury condo construction are matched by a retrenchment in public housing provisions; a nearby Wall Street boom generated seven- and eight-figure salaries while unemployment rose among the unskilled; poverty is increasingly concentrated among women, Latinos and African-Americans while social services are axed; and the conservatism of the 1980s spewed a recrudescence of racist violence throughout the city. With the emergence of deep recession in the early 1990s, rents have stabilized, but unemployment has soared. In the late 1990s the resurgence of gentrification and development is destined to magnify the polarization of the 1980s.

Tompkins Square lies deep in the heart of the Lower East Side. On its southern edge along Seventh Street a long slab of residential buildings overlooks the park, mostly late-nineteenth-century five- and six-storey walk-up tenements adorned with precariously affixed fire escapes, but also including a larger building with a dreary, modern, off-white facade. To the west, the tenements along Avenue A are barely more interesting, but many cross streets and the mix of smoke shops, Ukrainian and Polish restaurants, upscale cafes and hip bars, groceries, candy stores and night clubs make this the liveliest side of the park. Along Tenth Street on the northern edge stands a stately row of 1840s and 1850s townhouses, gentrified as far back as the early 1970s. To the east, Avenue B presents a more broken frontage: tenements, St. Brigid's Church from the mid–nineteenth century, and the infamous Christodora building – a sixteen-storey brick monolith built in 1928 that dominates the local skyline.

"One day," laments the tony, habitually understated AIA guide to New York architecture, "when this area is rebuilt, the mature park will be a godsend" (Willensky and White 1988: 163). Actually, the park itself is rather unexceptional. An oval rosette of curving, crisscross walkways, it is shaded by large plane trees and a few surviving elms. The walkways were lined by long rows of cement benches, replaced in the park reconstruction by wooden benches sectioned into individual seats by wrought iron bars designed to prevent homeless people from sleeping. Wide grassy patches, often bare, made up the body of the park and these were fenced off in the reconstruction. At the north end of the park are handball and basketball courts, playgrounds and the dog run, and at the south end a bandshell, which hosted everyone from the Fugs to the Grateful Dead in the 1960s to May Day demonstrations and the annual Wigstock Parade in the late 1980s. By day, before its reconstruction, the park would be filled with Ukrainian men playing chess, young guys selling drugs, yuppies walking to and from work, a few remaining punks with boom boxes, Puerto Rican women strolling babies, residents walking dogs, kids in the playgrounds. After 1988, there were also cops in cruisers, and photographers, and a growing population of evictees attracted to the relative safety of this "liberated" if still contested space. The encampments burgeoned before June 1991, and were made from tents, cardboard, wood, bright blue tarpaulins, and all sorts of scavenged material that could provide shelter. Hard drug users traditionally congregated in "crack alley" on the southern edge; a group of mostly working people clustered to the east, and Jamaican Rastafarians hung out by the temperance fountain closer to Avenue A. Political activists and squatters congregated closer to the bandshell, which also provided shelter during the rain. The bandshell was demolished in the reconstruction.

Variously scruffy and relaxing, free-flowing and energetic, but rarely dangerous unless the police are on maneuvers, Tompkins Square exemplifies the kind of neighborhood park that Jane Jacobs adopted as a *cause célèbre* in her famous antimodernist tract, *The Death and Life of Great American Cities* (1961). If it hardly has the physical features of a frontier, neither class conflict nor police riots are new to Tompkins Square Park. Originally a swampy "wilderness," its first evictees may have been the Manhattoes whose acceptance of some rags and beads in 1626 led to their loss of Manhattan Island. Donated to the city by the fur trader and capitalist John Jacob Astor, the swamp was drained, a park was constructed in 1834, and it was named after Daniel Tompkins, an ex-governor of New York State and US vice-president from 1817 to 1825. Immediately the park became a traditional venue for mass meetings of workers and the unemployed, although, to the apparent consternation of the populace, it was commandeered for use as a military parade ground in the 1850s and throughout the Civil War.

The symbolic power of the park as a space of resistance crystallized after 1873 when a catastrophic financial collapse threw unprecedented numbers of workers and families out of job and home. The city's charitable institutions were overwhelmed and at the urging of the business classes the city government refused to provide relief. "There was in any case a strong ideological objection to the concept of relief itself and a belief that the rigors of unemployment were a necessary and salutary discipline for the working classes" (Slotkin 1985: 338). A protest march was organized for January 13, 1874 in Tompkins Square, and the following account is reconstructed by labor historian Philip Foner:

> By the time the first marchers entered the Square, New Yorkers were witnessing the largest labor demonstration ever held in the city. The Mayor, who was expected to address the demonstration, changed his mind and, at the last minute, the police prohibited the meeting. No warning, however, had been given to the workers, and the men, women and children marched to Tompkins Square expecting to hear mayor Havemeyer present a program for the relief of the unemployed. When the demonstrators had filled the Square they were attacked by the police. "Police clubs," went one account, "rose and fell. Women and children went screaming in all directions. Many of them were trampled underfoot in the stampede for the gates. In the street bystanders were ridden down and mercilessly clubbed by mounted officers."
>
> (Foner 1978: 448)

Within an hour of the first baton charges, a special edition of the *New York Graphic* appeared in the streets with the headline: "A Riot Is Now in Progress in Tompkins Square Park" (Gutman 1965: 55).

Following the police riot the New York press provided a script that would have gratified the 1988 mayor. Decrying the marchers as "communists," and evoking the "red spectre of the commune," the New York *World* consistently built an analogy between the repression of the urban hordes in Tompkins Square and Colonel Custer's heroic Black Hills expedition against the savage Sioux of South Dakota. What began in 1874 as an outlandish juxtaposition between the park and the frontier (Slotkin 1985) had by the 1980s become an evocative but seemingly natural description.

The destiny of the Lower East Side has always been bound up with international events. The immigration of hundreds of thousands of European workers and peasants in the following decades only intensified the political struggles in the Lower East Side and its depiction in the press as a depraved environment. By 1910, some 540,000 people were crammed into the area's tenements, all competing for work and homes: garment workers, dockers, printers, laborers, craftsmen, shopkeepers, servants, public workers, writers, and a vital ferment of communists, Trotskyists, anarchists, suffragists and activist intellectuals devoted to politics and struggle. Successive economic recessions forced many into unemployment; tyrannical bosses, dangerous work conditions and a lack of workers' rights elicited large-scale union organizing. And landlords proved ever adept at rent gouging. The decade that began with the Triangle fire of 1911 – the fire engulfed 146 women garment workers from the Lower East Side, imprisoned behind locked sweatshop doors, forcing them to jump to their death in the street below – ended with the Palmer Raids of 1919 in which a wave of state-sponsored political terror was unleashed against the now notorious Lower East Side. In the 1920s as the suburbs burgeoned, landlords throughout the neighborhood allowed their buildings to fall into dilapidation, and many residents who could were following capital out to the suburbs.

Like other parks, Tompkins Square came to be viewed by middle-class reformers as a necessary "escape valve" for this dense settlement and volatile social environment. Following the 1874 riot, it was redesigned explicitly to create a more easily controllable space, and in the last decade of the century the reform and temperance movements constructed a playground and a fountain. The contest for the park ebbed and flowed, but took another surge during the Depression when Robert Moses redesigned the park, and again two decades later when the Parks Department tried unsuccessfully to usurp park land with a baseball diamond. Local demonstrations diverted this redesign (Reaven and Houck 1994). A hangout for Beat poets in the 1950s and the so-called counterculture in the 1960s, the park and its surroundings were again the scene of battles in 1967 when police waded into hippies sprawled out in the park in defiance of the "Keep off the Grass" signs.

This explosive history of the park belies its unremarkable form, making it a fitting locale for a "last stand" against gentrification.

BUILDING THE FRONTIER MYTH

Roland Barthes once proposed that "myth is constituted by the loss of the historical quality of things" (Barthes 1972: 129). Richard Slotkin elaborates that in addition to wrenching meaning from its historical context, myth has a reciprocal effect on history: "history becomes a cliché" (Slotkin 1985: 16, 21–32). We should add the corollary that myth is constituted by the loss of the *geographical* quality of things as well. Deterritorialization is equally central to mythmaking, and the more events are wrenched from their constitutive geographies, the more powerful the mythology. Geography too becomes a cliché.

The social meaning of gentrification is increasingly constructed through the vocabulary of the frontier myth, and at first glance this appropriation of language and landscape might seem simply playful, innocent. Newspapers habitually extol the courage of urban "homesteaders," the adventurous spirit and rugged individualism of the new settlers, brave "urban pioneers," presumably going where, in the words of *Star Trek*, no (white) man has ever gone before. "We find a place on the lower [*sic*] East Side," confesses one suburban couple in the genteel pages of the *New Yorker*:

> Ludlow Street. No one we know would think of living here. No one we know has ever heard of Ludlow Street. Maybe someday this neighborhood will be the way the Village was before we knew anything about New York. . . . We explain that moving down here is a kind of urban pioneering, and tell [Mother] she should be proud. We liken our crossing Houston Street to pioneers crossing the Rockies.
>
> ("Ludlow Street" 1988)

In its real estate section, the *New York Times* (March 27, 1983) announces "The Taming of the Wild Wild West," pursuant to the construction of the "Armory Condominium" two blocks west of Times Square:

> The trailblazers have done their work: West 42nd Street has been tamed, domesticated and polished into the most exciting, freshest, most energetic new neighborhood in all of New York . . . for really savvy buyers, there's the rapid escalation of land prices along the western corridor of 42nd Street. (After all, if the real estate people don't know when a neighborhood is about to bust loose, who does?)

As new frontier, the gentrifying city since the 1980s has been oozing with optimism. Hostile landscapes are regenerated, cleansed, reinfused with middle-class sensibility; real estate values soar; yuppies consume; elite gentility is democratized in mass-produced styles of distinction. So what's not to like? The contradictions of the actual frontier are not entirely eradicated in this imagery but they are smoothed into an acceptable groove. As with the Old West, the frontier is idyllic yet also dangerous, romantic but also ruthless. From *Crocodile Dundee* to *Bright Lights, Big City*, there is an entire cinematic genre that makes of urban life a cowboy fable replete with dangerous environment, hostile natives and self-discovery at the margins of civilization. In taming the urban wilderness, the cowboy gets the girl but also finds and tames his inner self for the first time. In the final scene of *Crocodile Dundee*, Paul Hogan accepts New York – and New York him – as he clambers like an Aussie sheepdog over the heads and shoulders of a subway crowd. Michael J. Fox can hardly end his fable by riding off into a reassuring western sunset since in the big city the bright lights are everywhere, but he does see a bright new day rise over the Hudson River and Manhattan's reconstructed financial district. The manifest destiny of the earlier frontier rains a reciprocal Valhalla on the big city.

The frontier myth of the new city is here so clichéd, the geographical and historical quality of things so lost, that we may not even see the blend of myth in the landscape. This merely testifies to the power of the myth, but it was not

Plate 1.4 Real estate capital rides the new urban frontier

always so. The analogy between the 1874 Tompkins Square marchers and the Sioux Nation was at best tentative and oblique, the mythology too young to bear the full ideological weight of uniting such obviously disparate worlds. But the real and conceptual distance between New York and the Wild Wild West has been continually eroded; perhaps the most iconoclastic evocation of a frontier in the early city came only a few years after Custer's Black Hills campaign when a stark, elegant but isolated residential building rose in the boonies of Central Park West and was named "The Dakota Apartments." By contrast, in the condomania that has engulfed Manhattan a century later – an environment in which any social, physical or geographical connection with the earlier frontier is obliterated – the "Montana," "Colorado," "Savannah" and "New West" have been shoehorned into already overbuilt sites with ne'er a comment about any iconographic inconsistency. As history and geography went west, the myth settled east, but it took time for the myth itself to be domesticated into the urban environment.

The new urban frontier motif encodes not only the physical transformation of the built environment and the reinscription of urban space in terms of class and race, but also a larger semiotics. Frontier is a style as much as a place, and the 1980s saw the faddishness of Tex-Mex restaurants, the ubiquity of desert decor, and a rage for cowboy chic, all woven into the same urban landscapes of consumption. A *New York Times* Sunday Magazine clothing advertisement (August 6, 1989) gives the full effect:

> For urban cowboys a little frontier goes a long way. From bandannas to boots, flourishes are what counts. . . . The Western imprint on fashion is now much like a cattle brand – not too striking, but obvious enough to catch the eye. For city dudes, that means accents: a fringed jacket with black leggings; a shearling coat with a pin-stripe suit; a pair of lizard boots with almost anything. When in doubt about the mix stride up to the mirror. If you're inclined to say "Yup," you've gone too far.

New York's upmarket boutiques dispensing fashionable frontier kitsch are concentrated in SoHo, an area of artists' lofts and effete galleries, gentrified in the late 1960s and 1970s, and enjoying an unprecedented boom in the 1980s. SoHo borders the Lower East Side to the west and southwest. Here, "frontier" aspires on occasion to philosophy. Zona, on Greene Street, sells Navajo rugs, "Otomi Indian natural bark notepaper," Santa Fe jewelry, terra-cotta pottery, "Lombak baskets in rich harvest colors," bola ties. Zona oozes authenticity. All the "pieces" are numbered and a catalogue of the "collection" has been produced. On a small, plain, deliberately understated sign, with writing embossed on gold paper, the store offers its "personal" philosophy of craft-friendliness suffused with more than a whiff of New Age spiritualism:

> At a time when the ever expanding presence of electronic tools and high technology is so pervasive the need to balance our lives with products that celebrate the textual and sensorial become essential. We think of our customers as resources and not simply as consumers. We are guided by the belief that information is energy and change is the constant.
> Thank you for visiting our space.

Americana West, on Wooster Street, strives for a purer desert look. On the sidewalk outside the front door, a patrician Indian chief complete with tomahawk and feathered headgear stands guard. The window display features a bleached buffalo skull for $500 while inside the store are sofas and chairs made from longhorns and cattle skin. A gallery as much as a store, Americana West purveys diverse images of noble savages, desert scenes à la Georgia O'Keeffe, petroglyphs and pictographs, whips and spurs. Cacti and coyotes are everywhere (none real); a neon prickly pear is available for $350. In lettering on the front window, Americana West announces its own theme, a crossover cultural geography between city and desert: "The Evolving Look of the Southwest. Designers Welcome . . . Not for City Slickers Only."

The frontier is not always American nor indeed male. At La Rue des Rêves the theme is jungle eclectic. Leopard coats (faux of course), antelope leather skirts, and chamois blouses seem still alive, slinking off their hangers toward the cash registers. Fashion accessories dangle like lianas from the jungle canopy. A stuffed gorilla and several live parrots round out the ambience. La Rue des Rêves may have been "too, too" – it was a casualty of the late 1980s stock market crash – but the theme has survived in clothing chains as well as boutiques. At the Banana Republic customers have their safari purchases packed in brown paper bags sporting a rhinoceros. On the silver screen, meanwhile, movies such as *Out of Africa* and *Gorillas in the Mist* reinforce the vision of pioneering whites in darkest Africa, but with heroines for heroes. As middle-class white women come to play a significant role in gentrification their prominence on earlier frontiers is rediscovered and reinvented. Thus designer Ralph Lauren began the 1990s with a collection centered on "the Safari woman." He explains thus the romantic and nostalgic ur-environmentalism that drove him to it: "I believe that a lot of wonderful things are disappearing from the present, and we have to take care of them." A mahogany four-poster draped in embroidered mosquito netting, jodhpurs, faux ivory, and a "Zanzibar" bedroom set patterned with Zebra stripes surround Lauren's "Safari Woman," herself presumably an endangered species. Originally Ralph Lifschitz born in the Bronx, but now ensconced on a Colorado ranch half the size of that borough, "Lauren" has never been to Africa – "sometimes it's better if you haven't been there" – but feels well able to represent it in and for our urban fantasies. "I'm trying to evoke a world in which there was this graciousness we could touch. Don't look at yesterday. We can have it. Do you want to make the movie you saw a reality? Here it is" (Brown 1990).

Even as Africa is underdeveloped by international capital, engulfed by famine and wars, it is remarketed in Western consumer fantasies – but as the preserve of privileged and endangered whites. As one reviewer put it, the safari collection "smacks of bwana style, of Rhodesia rather than Zimbabwe" (Brown 1990). Lauren's Africa is a country retreat for and from the gentrified city. It provides the decorative utensils by which the city is reclaimed from wilderness and remapped for white upper-class settlers with global fantasies of again owning the world – recolonizing it from the neighborhood out.

Nature too is rescripted on the urban frontier. The frontier myth – originally engendered as an historicization of nature – is now reapplied as a naturalization of urban history. Even as rapacious economic expansion destroys deserts and rain forests, the new urban frontier is nature-friendly: "All woods used in [Lauren's Safari] collection are grown in the Philippines and are not endangered" (Brown 1990). The Nature Company, a chain store with a branch in South Street Seaport at the south end of the Lower East Side, is the apotheosis of this naturalized urban history, selling maps and globes, whaling anthologies and telescopes, books on dangerous reptiles, and stories of exploration and conquest. The store's unabashed nature idolatry and studied avoidance of anything urban are the perfect disappearing mirror in which contested urban histories are refracted (N. Smith 1996b). In affirming the connection with nature, the new urban frontier erases the social histories, struggles and geographies that made it.

The nineteenth century and its associated ideology were "generated by the social conflicts that attended the 'modernization' of the Western nations," according to Slotkin. They are "founded on the desire to avoid recognition of the perilous consequences of capitalist development in the New World, and they represent a displacement or deflection of social conflict into the world of myth" (Slotkin 1985: 33, 47). The frontier was conveyed in the city as a safety valve for the urban class warfare brewing in such events as the 1863 New York draft riot, the 1877 railway strike, and indeed the Tompkins Square riot of 1874. "Spectacular violence" on the frontier, Slotkin concludes, had a redemptive effect on the city; it was "the alternative to some form of civil class war which, if allowed to break out within the metropolis, would bring about a secular *Götterdämmerung*" (Slotkin 1985: 375). Projected in press accounts as extreme but comparable versions of events in the city, a magnifying mirror to the most ungodly depravity of the urban masses, reportage of the frontier posited eastern cities as a paradigm of social unity and harmony in the face of external threat. Urban social conflict was not so much denied as externalized, and whosoever disrupted this reigning urban harmony committed unnatural acts inviting comparison with the external enemy.

Today the frontier ideology continues to displace social conflict into the realm of myth, and at the same time to reaffirm a set of class-specific and race-specific social norms. As one respected academic has proposed, unwittingly replicating Turner's vision (to not a murmur of dissent), gentrifying neighborhoods should be seen as combining a "civil class" who recognize that "the neighborhood good is enhanced by submitting to social norms," and an "uncivil class" whose behavior and attitudes reflect "no acceptance of norms beyond those imperfectly specified by civil and criminal law." Neighborhoods might then be classified "by the extent to which civil or uncivil behavior dominates" (Clay 1979a: 37–38).

The frontier imagery is neither merely decorative nor innocent, therefore, but carries considerable ideological weight. Insofar as gentrification infects working-class communities, displaces poor households, and converts whole neighborhoods into bourgeois enclaves, the frontier ideology rationalizes social differentiation and exclusion as natural, inevitable. The poor and working class are all too easily defined as "uncivil," on the wrong side of a heroic dividing line, as savages and communists. The substance and consequence of the frontier imagery is to tame the wild city, to socialize a wholly new and therefore challenging set of processes into safe ideological focus. As such, the frontier ideology justifies monstrous incivility in the heart of the city.

SELLING LOISAIDA

The frontier takes different forms in different places; it adapts to place as it makes place. But everywhere the frontier line is variously present. A *Wall Street Journal* reporter describes dining possibilities in "Indian Country" at the end of the 1980s: "For dining a new restaurant on Avenue C called 'Bernard' offers 'organic French cuisine.' Frosted glass windows protect diners from the sight of the burned out tenements across the street as they nibble their $18 loins of veal" (Rickelfs 1988). Shades of Baudelaire in Haussmann's Paris, as we shall see. Notice that the poor, abandoned and homeless of the neighborhood were already invisible without the frosted window; only the burned out shells from which they were evicted threaten to intrude.

On the Lower East Side two industries defined the new urban frontier that emerged in the 1980s. Indispensable, of course, is the real estate industry which christened the northern part of the Lower East Side the "East Village" in order to capitalize on its geographical proximity to the respectability, security, culture and high rents of Greenwich Village. Then there is the culture industry – art dealers and patrons, gallery owners and artists, designers and critics, writers and performers – which has converted urban dilapidation into ultra chic. Together in the 1980s the culture and real estate industries invaded this rump of Manhattan from the west. Gentrification and art came hand in hand, "slouching toward Avenue D," as art critics Walter Robinson and Carlo McCormick (1984) put it. Block by block, building by building, the area was converted to a landscape of glamour and chic spiced with just a hint of danger.

The rawness of the neighborhood has in fact been part of the appeal. Only in the Lower East Side have art critics celebrated "minifestivals of the slum arts"; only here have artists cherished "a basic ghetto material – the ubiquitous brick"; and only here would the art entourage blithely admit to being "captivated by the liveliness of ghetto culture" (Moufarrege 1982, 1984). Alongside the gallery called "Fun," the knickknack boutique named "Love Saves the Day," and the bar called "Beulah Land" (Bunyan's land of rest and quiet) came "Civilian Warfare" and "Virtual Garrison" (both galleries), "Downtown Beirut" (a bar) and an art showing called "The Twilight Zone." Frontier danger permeated the very art itself, whatever the nostalgic eclecticism of the Lower East Side scene. The "law of the jungle" ruled the new art scene, an art scene driven by "savage energy," gushed Robinson and McCormick (1984: 138, 156). Neoprimitivist art, in fact, depicting black-figured urban "natives," often running wild in the streets, was a central theme of this "savage energy."

The most insightful critique of this connection between art and real estate remains that by Rosalyn Deutsche and Cara Ryan in a classic article, "The fine art of gentrification" (Deutsche and Ryan 1984). The complicity of art with gentrification is no mere serendipity, they show, but "has been constructed with the aid of the entire apparatus of the art establishment." Linking the rise of the "East Village" with the triumph of neo-Expressionism in art, they argue that however countercultural its pose, the broad abstention from political self-reflection condemned Lower East Side art to reproducing the dominant culture. The unprecedented commodification of art in the 1980s engendered an equally ubiquitous aestheticization of culture and politics: graffiti came off the trains and into the galleries, while the most outrageous punk and new-wave styles moved rapidly from the streets to full-page advertisements in the *New York Times*. The press began sporting stories about the opulence of the new art scene – at least for some: Don't let the poverty of the Lower East Side fool you, was the message; this generation of young artists gets by with American Express Goldcards (Bernstein 1990).

The simultaneous disavowal of social and political context and dependence on the cultural establishment placed avant-garde artists in a sharply contradictory position. They came to function as "broker" between the culture industry and the majority of still-aspiring artists. Lower East Side galleries played the pivotal role: they provided the meeting place for grassroots ambition and talent and establishment money (Owens 1984: 162–163).[2] Representing and patronizing the neighborhood as a cultural mecca, the culture industry attracted tourists, consumers, gallery gazers, art patrons, potential immigrants – all fueling gentrification. Not all artists so readily attach themselves to the culture establishment, of course, and a significant artists' opposition survived the commodification and price escalation that boosted the neighborhood's twin industries in the 1980s. Following the Tompkins Square riot, in fact, there was a flourishing of political art aimed squarely at gentrification, the police and the art industry. Some artists were also squatters and housing activists, and a lot of subversive art was displayed as posters, sculpture and graffiti in the streets or in more marginal gallery spaces (see for example Castrucci *et al.* 1992).

For the real estate industry, art tamed the neighborhood, refracting back a mock pretense of exotic but benign danger. It depicted the East Village as rising from low life to high brow. Art donates a salable neighborhood "personality," packaged the area as a real estate commodity and established demand. Indeed, "the story of the East Village's newest bohemian efflorescence," it has been suggested, "can also be read as an episode in New York's real estate history – that is, as the deployment of a force of gentrifying artists in lower Manhattan's last slum" (Robinson and McCormick 1984: 135).

By 1987, however, the marriage of convenience between art and real estate started to sour, and a wave of gallery closures was precipitated by massive rent increases demanded by landlords unconstrained by rent control. It is widely speculated that these landlords – many of them anonymous management companies operating out of post office boxes – offered artificially low rents in the early 1980s in order to attract galleries and artists whose presence would hype the area and hike rents. Handsomely successful, they demanded sharp increases as the first five-year leases came due. The neighborhood was now saturated with as many as seventy galleries, artistic and economic competition was cutthroat, and a financial shakeout, synchronized with the 1987 stock market crash, ensued. First Avenue was manifestly not "downtown Beirut" and a host of artistic and financial fantasies plummeted to earth. Many galleries closed. The most successful decamped to SoHo where gentrification capital also regrouped; the less successful (financially) often went across the bridge to Williamsburg in Brooklyn. Left in the lurch by the real estate industry, many Lower East Side artists were also summarily dropped by a cultural elite that had found other dalliances (Bernstein 1990) – but not before the culture industry as a whole had spearheaded a fundamental shift in the neighborhood's image and real estate market.

That some artists became victims of the very gentrification process they helped precipitate, and that others actively opposed the process, has touched off a debate in the art press (Owens 1984: 162–163; Deutsche and Ryan 1984: 104; Bowler and McBurney 1989). However wittingly or otherwise, the culture and real estate industries worked together to transform the Lower East Side into a new place – different, unique, a phenomenon, the pinnacle of avant-garde fashion. Fashion and faddishness created cultural scarcity much as the real estate industry's demarcation of the "East Village" instantaneously establishes a scarcity of privileged addresses. Good art and good locations become fused. And good location means money.

PIONEERING FOR PROFIT

The Lower East Side has experienced several phases of rapid building associated with larger economic cycles, and the present-day built environment results from this history. A few early buildings remain from the 1820s to 1840s, but rectangular "railroad" tenements are more common, built in the 1850s through the Civil War to house the largely immigrant working class. These are the tenements that figured so vividly in Jacob Riis's 1901 *How the Other Half Lives* (Riis 1971 edn.). In the decade and a half after 1877, with the economy expanding and immigration growing, the area experienced its most intense building boom. Virtually all vacant land was developed with "dumb-bell" tenements, so named because the rectangular form of the traditional railway tenements were now forced by law to include dumbbell-shaped airshafts between structures. By the 1893 economic crash, which effectively ended this building boom, almost 60 percent of all New York City housing comprised dumbbell tenements; at least 30,000 such buildings throughout the city are still inhabited, with the largest concentration in the Lower East Side (G. Wright 1981: 123). The next building boom, beginning in 1898, was concentrated at the urban edge; the Lower East Side did receive some "new law" tenements (post-1901, when a new law required improved design standards), but many landlords in the area had already begun disinvesting, neglecting maintenance and repairs on their grossly overcrowded buildings.

New York's ruling class has long sought to tame and reclaim the Lower East Side from its unruly working-class hordes. Only five years after the federal government severely curtailed European immigration, the Rockefeller-sponsored Regional Plan Association offered an extraordinary vision for the Lower East Side. The 1929 New York Regional Plan explicitly envisaged the removal of the existing population, the reconstruction of "high-class residences," modern shops, a yacht marina on the East River, and the physical redevelopment of the Lower East Side highway system in such a way as to strengthen the connection with neighboring Wall Street:

> The moment an operation of this magnitude and character was started in a district, no matter how squalid it was, an improvement in quality would immediately begin in adjacent property and would spread in all directions. New stores would start up prepared to cater to a new class of customers. The streets thereabouts would be made cleaner. Property values would rise. . . . After a while, other apartment units would appear and in the course of time the character of the East Side would be entirely changed.
>
> (quoted in Gottlieb 1982; see also Fitch 1993)

The stock market crash of 1929, the ensuing Depression and World War II, the unprecedented wave of postwar suburban expansion, and eventually the New York City fiscal crisis all mitigated against the planned reinvestment and reconstruction of the Lower East Side as a high-class haven. Various slum clearance and low-income residential projects were initiated between the late 1930s and early 1960s, but, combined with the withdrawal of capital, these policies often intensified the long-term economic and social processes laying waste to the Lower East Side and other such neighborhoods. In the postwar period, disinvestment and abandonment, demolition and public warehousing, were the major tactics of a virulent antiurbanism that converted the Lower East Side into something of a free-fire zone. Especially hard hit was the area south of Houston Street and the Alphabet City area to the east between Avenue A and Avenue D. Urban renewal here simply reinforced the ghettoization of poor residents, especially Latinos, amid the rubble of disinvestment.

Not until a further half-century of disinvestment, dilapidation and decline did the 1929 vision begin to be implemented. Even as yuppies and artists began to pick over the wreckage in the late 1970s, everyone else was moving out. From the 1910 peak population of over half a million, the Lower East Side lost almost 400,000 inhabitants over the next seven decades; in the 1970s it lost 30,000, giving it a 1980 population of nearly 155,000. In the heart of Loisaida between Houston and Tenth Streets, Avenue B to Avenue D – so-called Alphabet City – where abandonment and property disinvestment have been most intense, the population declined by an extraordinary 67.3 percent. The median household income of $8,782 was only 63 percent of the 1980 citywide figure, and twenty-three of twenty-nine census tracts in the area experienced an increase in the number of families living below the poverty level. In Alphabet City it was the poor who were left behind; 59 percent of the remaining population survived below the poverty level. The neighborhood so deliberately colonized by yuppies and artists at the end of the 1970s was the poorest in Manhattan outside Harlem. In the 1980s, the neighborhood actually experienced a population reversal with 161,617 recorded in the 1990 census.

Declining property values accompanied declining populations in the 1970s and much of the 1980s. Consider the case of 270 East 10th Street, a run-down but occupied five-storey dumbbell tenement between First Avenue and Avenue A, half a block west of Tompkins Square Park. In 1976, at the time of peak disinvestment, it was sold by a landlord who simply wanted out; the price was a mere $5,706 plus the assumption of unpaid property taxes. By the beginning of 1980 it was resold for $40,000. Eighteen months later it went for $130,000. In September 1981 the building was sold again, this time to a New Jersey real estate concern for $202,600. In less than two years the building's price multiplied five times – without any renovation (Gottlieb 1982).

This is not an unusual case. On Tompkins Square Park the sixteen-storey Christodora Building, now a symbol of antigentrification struggle, experienced a similar cycle of disinvestment and reinvestment. Built in 1928 as a settlement house, the Christodora was sold to the City of New York in 1947 for $1.3 million. It was used for various City functions and eventually as a community center and hostel, housing among others the Black Panthers and the Young Lords. Run down and dilapidated by the late 1960s, the building attracted no bids at a 1975 auction. It was later sold for $62,500 to a Brooklyn developer, George Jaffee. The doors of the deserted building had been welded shut and remained that way for five years while Jaffee unsuccessfully sought federal funds for rehabilitating the Christodora as low-income housing. In 1980 Jaffee began to get inquiries about the building. The welder was called to provide entry, the building was inspected, and offers of $200,000 to $800,000 began to materialize. Jaffee eventually sold the building in 1983 for $1.3 million to another developer, Harry Skydell, who in turn "flipped" the building a year later for $3 million, only to recoup it later in a joint venture with developer Samuel Glasser. Skydell and Glasser renovated the Christodora and in 1986 marketed its eighty-six condominium apartments. The quadruplex penthouse, with private elevator, three terraces and two fireplaces, was offered for sale in 1987 for $1.2 million (Unger 1984; DePalma 1988).

At 270 East Tenth, at the Christodora, and at hundreds of other buildings in the Lower East Side, it is real estate profits, first and foremost, that are revitalized. The Tompkins Court, a 1988 rehabilitation, offered one-bedroom units at the "post-87 crash bargain price" of $139,000–$209,000, two-bedroom units for $239,000–$329,000. For the least expensive of these an estimated annual household income of $65,000 was required; for the most expensive an income of $160,000. Even the small studios were inaccessible to those earning less than $40,000. Several blocks away at another tenement rehab, seventeen co-ops were sold, with two-bedroom units ranging from $235,000 to $497,800 (Shaman 1988). Mortgage and maintenance costs on the latter amounted to almost $5,000 per month. Two months' payment on this apartment exceeded the neighborhood's median annual income. Only by the early 1990s did sale prices begin to drop appreciably – as much as 15–25 percent at the top end of the market but less at lower rental levels.

Unrestrained by rent control of any sort, commercial rents and sales rose even faster. Long-time small businesses were forced out as landlords indiscriminately raised rents. Maria Pidhorodecky's Italian-Ukrainian restaurant, the Orchidia, a fixture on Second Avenue since 1957, closed in the mid-1980s when the landlord was able to raise the rent for the 700-square-foot space from $950 to $5,000 (Unger 1984).

In his investigation of the workings of the Lower East Side real estate market, journalist Martin Gottlieb uncovered the results of the rent gap (see chapter 3) first hand. At 270 East Tenth Street, for example, while the combined sale price of building and land soared from $5,706 to $202,600 in five and a half years, the value of the building alone, according to city property tax assessors, actually fell from $26,000 to $18,000. And this is a typical result; even taking into account the structured undervaluation of buildings *vis-à-vis* the market, the land is much more valuable than the building. The perverse rationality of real estate capitalism means that building owners and developers garner a double reward for milking properties and destroying buildings. First, they pocket the money that should have gone to repairs and upkeep; second, having effectively destroyed the building and established a rent gap, they have produced for themselves the conditions and opportunity for a whole new round of capital reinvestment. Having produced a scarcity of capital in the name of profit they now flood the neighborhood for the same purpose, portraying themselves all along as civic-minded heroes, pioneers taking a risk where no one else would venture, builders of a new city for the worthy populace. In Gottlieb's words, this self-induced reversal in the market means that a "Lower East Side landlord can drink his milk and have it too" (Gottlieb 1982).

The economic geography of gentrification is not random; developers do not just plunge into the heart of slum opportunity, but tend to take it piece by piece. Rugged pioneersmanship is tempered by financial caution. Developers have a vivid block-by-block sense of where the frontier lies. They move in from the outskirts, building "a few strategically placed outposts of luxury," as Henwood (1988: 10) has put it. They "pioneer" first on the gold coast between safe neighborhoods on one side where property values are high and the disinvested slums on the other where opportunity is higher. Successive beachheads and defensible borders are established on the frontier. In this way economic geography charts the strategy of urban pioneering.

Whereas the myth of the frontier is an invention that rationalizes the violence of gentrification and displacement, the everyday frontier on which the myth is hung is the stark product of entrepreneurial exploitation. Thus whatever its visceral social and cultural reality, the frontier language camouflages a raw economic reality. Areas that were once sharply redlined by banks and other financial institutions were sharply "greenlined" in the 1980s. Loan officers are instructed to take down their old maps with red lines around working-class and minority neighborhoods and replace them with new maps sporting green lines: make every possible loan within the greenlined neighborhood. In the Lower East Side as elsewhere, the new urban frontier is a frontier of profitability. Whatever else is revitalized, the profit rate in gentrifying neighborhoods is revitalized; indeed many working class neighborhoods experience a dramatic "devitalization" as incoming yuppies erect metal bars on their doors and windows, disavow the streets for parlor living, fence off their stoops, and evict undesirables from "their" parks.

If the real estate cowboys invading the Lower East Side in the 1980s used art to paint their economic quest in romantic hues, they also enlisted the cavalry of city government for more prosaic tasks: reclaiming the land and quelling the natives. In its housing policy, drug crackdowns, and especially in its parks strategy, the City devoted its efforts not toward providing basic services and living opportunities for existing residents but toward routing many of the locals and subsidizing opportunities for real estate development. A 1982 consultants' report entitled *An Analysis of Investment Opportunities in the East Village* captured the City's strategy precisely: "The city has now given clear signals that it is prepared to aid the return of the middle class by auctioning city-owned properties and sponsoring projects in gentrifying areas to bolster its tax base and aid the revitalization process" (Oreo Construction Services 1982).

The City's major resource was its stock of "*in rem*" properties, mostly fore-closed from private landlords for nonpayment of property taxes. By the early 1980s the Department of Housing, Preservation and Development held over 200 such *in rem* buildings in the Lower East Side and a similar number of vacant lots. With sixteen of these properties, the Koch administration made its first significant foray into the real estate frenzy of gentrification; artists were to be the vehicle. In August 1981 HPD solicited proposals for an Artist Homeownership Program (AHOP) and the next year announced a renovation project that was to yield 120 housing units in sixteen buildings, each costing an estimated $50,000, aimed at artists earning at least $24,000. Their purpose, the Mayor proclaimed, was "to renew the strength and vitality of the community," and five artists' groups and two developers were selected to execute the $7 million program (Bennetts 1982).

But many in the community disagreed vigorously enough to oppose the AHOP plan. The Joint Planning Council, a coalition of more than thirty Loisaida housing and community organizations, demanded that so valuable a resource as abandoned buildings should be renovated for local consumption; city councilwoman Miriam Friedlander saw the plan as "just a front for gentrification"; "the real people who will profit from this housing are the developers who renovate it." And indeed, the HPD Commissioner expressed the fervent hope that the project would be "a stimulus for overall neighbor-hood revitalization." While supporting artists portrayed themselves as normal folks, just part of the working class, a population already largely displaced from Manhattan who deserved housing as much as anyone else, an artists' opposition emerged – "Artists for Social Responsibility" – who opposed the use of artists to gentrify the neighborhood. HPD, the mayor and AHOP were ultimately defeated by the City Board of Estimates, which refused to provide the initial $2.4 million of public funds (Carroll 1983).

But AHOP was a warm-up for a larger auction program, as HPD prepared to leverage gentrification citywide using *in rem* properties. The Joint Planning Council decided to grab the initiative by proposing its own community-based plan, and in 1984 it proposed that all City-owned vacant lots and properties be used for low- and moderate-income housing and that the speculation responsible for eliminating existing low-income units be controlled. The City ignored the community plan and came back with a "cross-subsidy" program. HPD would sell City-owned properties to developers, either by auction or at appraised value, in return for an agreement by developers that a vaguely specified 20 percent of rehabilitated or newly built units would be reserved for tenants unable to afford market rates. Developers would receive a tax subsidy in return. Initially some community groups gave the program tentative support; others sought to adjust the ratio of market-rate to subsidized housing to 50:50, while others rejected the entire idea as a backdoor route to building minimal public housing.

But opposition mounted as the actual intent of the program became clear. In 1988 the City announced that the Lefrak Organization – a major national developer – would build on the Seward Park site where, in 1967, 1,800 poor people, mostly African-American and Latino, were displaced when their homes were urban renewed. They were promised the new apartments scheduled for the site, but twenty years later the renewal was yet to happen. The fee for the site was $1, and Lefrak would pay a further $1 per year for the ninety-nine-year lease. Under the plan, Lefrak would build 1,200 apartments, 400 of which would be market-rate condominiums, 640 would be rented at $800–$1,200 to "middle-income" households earning $25,000–$48,000, and the remaining 160 units would go as "moderate-income" units to those earning $15,000–$25,000. No apartments were actually earmarked for low-income people. Further, all rental units would revert to Lefrak as luxury co-ops on the open market after twenty years; Lefrak would get a thirty-two-year tax abatement, and an overall City subsidy of $20 million. Lawyers representing several of the 1967 tenants filed a class action suit against the Lefrak condo. "Yupper-income housing in low income neighborhoods" is how one housing advocate described the plan, "and the purpose is creating hot new real-estate markets" (Glazer 1988; Reiss 1988). The project got as far as a "Memorandum of Understanding" with the City, but as the depression closed in, the folly of attaching any subsidized housing to market development became clear. Lefrak abandoned the project – but not before it became clear that the City had no intention of mandating Lefrak to build the 20 percent of subsidized units in the same neighborhood. The geographical mobility of the subsidized housing of course opened up the specter of gentrification again for those who had not already seen through the "double-cross subsidy" program, as it came to be known by community activists.

With AHOP and the cross-subsidy proposal, the City led the economic cavalry charge into the Lower East Side, but it also resorted to a little mood creation. In an effort to clear the streets of "natives" who might hinder the gentrification frontier, Operation Pressure Point was launched in January 1984. An estimated 14,000 drug busts were made in eighteen months throughout the Lower East Side, and the *New York Times* gloated that "thanks to operation pressure point, art galleries are replacing shooting galleries." But the petty offenders were quickly released, the kingpins never apprehended, and when the pressure eased the street sellers returned.

Along with Operation Pressure Point, the City organized an assault on the parks as part of its wider gentrification strategy. As developer William

Zeckendorf Jr. secured massive tax abatements and zoning variations for his twenty-eight-storey luxury Zeckendorf Tower at Fourteenth Street and Broadway – a "fortress" development intended to anchor future forays into the Lower East Side – the City had already weighed in with tactical support.[3] The plan evicted the homeless and others of the "socially undesirable population" from adjacent Union Square Park, and began a two-year, $3.6 million renovation. Inaugurating the renovation in the spring of 1984, Mayor Koch justified the Zeckendorf subsidy by blaming the victims: "First the thugs took over, then the muggers took over, then the drug people took over, and now we are driving them out" (quoted in Carmody 1984). In its initial sparkling antisepsis, the new park complemented the facade of the Zeckendorf condo. Some trees have been thinned out, walls knocked down, paths widened and an open plaza constructed at the south end, all offering long-range visibility for surveillance and control. Sharp-edged, bright new stonework replaced slabs worn gray by weather and footsteps, the farmers' market was spruced up but retained, and the park's monuments cleaned and polished in a nostalgic "restoration" of a nonexistent past. The same strokes that deoxidized the park's green statues back to their gleaming bronze splendor attempted to wipe away the city's history of homelessness and poverty. As Rosalyn Deutsche concluded, "the aesthetic presentation of the physical site of development is indissolubly linked to the profit motives impelling Union Square's "revitalization" (Deutsche 1986: 80, 85–86).

If the gentrification of Union Square Park hardly lived up to expectations, with patrolling cops and returning evictees very much restoring the park to the frontier edge, the City nonetheless persevered. The City's efforts moved south to Washington Square Park in the Village, where, as in Union Square Park, boundary fences were erected, a curfew imposed, police patrols stepped up. Then in 1988 they moved east into Tompkins Square Park. Rebuffed by the summer demonstrations culminating in the August police riot, the City's traditional park gentrification strategy of curfews and closures followed by "restoration" was defeated – for a time – by the August riot.

"ANOTHER WAVE MORE SAVAGELY THAN THE FIRST":[4] THE NEW (GLOBAL) INDIAN WARS?

"A sort of wartime mentality seems to be settling onto New Yorkers affected by the housing squeeze," commented *New York* magazine as the gentrification boom got under way in the early 1980s (Wiseman 1983). Especially in the Lower East Side, the geography of recent urban change reveals the future gentrified city, a city sparkling with the neon of elite consumption anxiously cordoned off from homeless deprivation. As the gentrification frontier came to course through neighborhood after neighborhood, most rapidly during economic expansions, but rarely at a slouch, previously working class sections of the city were dragged into the international circuits of capital. While Lower East Side art was shown in London or Paris, the neighborhood's fanciest condos were advertised in *The Times* and *Le Monde*.

Gentrification portends a class conquest of the city. The new urban pioneers seek to scrub the city clean of its working-class geography and history. By remaking the geography of the city they simultaneously rewrite its social history as a preemptive justification for a new urban future. Slum tenements become historic brownstones, and exterior facades are sandblasted to reveal a future past. Likewise with interior renovation. "Inner worldly asceticism becomes public display" as "bare brick walls and exposed timbers come to signify cultural discernment, not the poverty of slums without plaster" (Jager 1986: 79–80, 83, 85). Physical effacement of original structures effaces social history and geography; if the past is not entirely demolished it is at least reinvented – its class and race contours rubbed smooth – in the refurbishment of a palatable past.

Where the militance or persistence of working-class communities or the extent of disinvestment and dilapidation would seem to render such genteel reconstruction a Sisyphean task, the classes can be juxtaposed by other means. Squalor, poverty and the violence of eviction are constituted as exquisite ambience. The rapid polarization of new classes in the making is glorified for its excitement rather than condemned for its violence or understood for the rage it threatens.

The effort to recolonize the city involves systematic eviction. In its various plans and task force reports for gentrifying what remains of the inner city, New York City government has never proposed a plan for relocating evictees. This is stunning testimony to the real program. Denying any connection between gentrification and displacement, City officials refuse to admit the possibility that gentrification causes homelessness. Public policy is geared to allow the housed to "see no homeless," in the words of one Lower East Side stencil artist. The 1929 Regional Plan for the Lower East Side was at least more honest:

> Each replacement will mean the disappearance of many of the old tenants and the coming in of other people who can afford the higher rentals required by modern construction on high priced land. Thus in time economic forces alone will bring about a change in the character of much of the East Side population.
>
> (quoted in Gottlieb 1982: 16)

One developer justifies the violence of the new frontier: "To hold us account-able for it is like blaming the developer of a high-rise building in Houston for the displacement of the Indians a hundred years before" (quoted in Unger 1984: 41). In Burlington, Vermont, one restaurateur has taken seriously the mission of getting "those people" out of sight. The owner of Leunig's Old World Cafe, in the gentrified, cobblestone, boutique-filled Church Street Marketplace, became incensed at the homeless people who, he said, were "terrorizing" his restaurant's clients. Funded by donations from restaurateurs and other local businessmen in the town, he began an organization called "Westward Ho!" to provide homeless people with one-way tickets out of town – to Portland, Oregon.

Some have gone further in the effort to see no homeless, hoping in fact to illegalize homelessness altogether:

If it is illegal to litter the streets, frankly it ought to be illegal . . . to sleep in the streets. Therefore, there is a simple matter of public order and hygiene in getting these people somewhere else. Not arrest them, but move them off somewhere where they are simply out of sight.

(George Will, quoted in Marcuse 1988: 70)

This kind of vengeful outburst only lends more weight to Friedrich Engels' famous admonition of more than a century ago:

the bourgeoisie has only one method of settling the housing question. . . . The breeding places of disease, the infamous holes and cellars in which the capitalist mode of production confines our workers night after night are not abolished; they are merely *shifted elsewhere*.

(Engels, 1975 edn., 71, 73–4; emphasis in original)

Evicted from the public as well as the private spaces of what is fast becoming a downtown bourgeois playground, minorities, the unemployed and the poorest of the working class are destined for large-scale displacement. Once isolated in central city enclaves, they are increasingly herded to reservations on the urban edge. New York's HPD becomes the new Department of the Interior; the Social Security Administration the new Bureau of Indian Affairs; and Latino, African-American and other minorities the new Indians. At the beginning of the onslaught, one especially prescient East Village developer was cynically blunt about what the new gentrification frontier would mean for evictees as gentrification raced toward Avenue D: "They'll all be forced out. They'll be pushed east to the river and given life preservers" (quoted in Gottlieb 1982: 13).

The dramatic shifts affecting gentrifying neighborhoods are experienced as intensely local. The Lower East Side is a world away from the upper-crust *noblesse* of the Upper East Side three miles north; and within the neighborhood, Avenue C is still a very different place from First Avenue. Yet the processes and forces shaping the new urbanism are global as much as local. Gentrification and homelessness in the new city are a particular microcosm of a new global order etched first and foremost by the rapacity of capital. Not only are broadly similar processes remaking cities around the world, but the world itself impinges dramatically on these localities. The gentrification frontier is also an "imperial frontier," says Kristin Koptiuch (1991: 87–89). Not only does international capital flood the real estate markets that fuel the process, but international migration provides a workforce for many of the professional and managerial jobs associated with the new urban economy – a workforce that needs a place to stay. Even more does international migration provide the service workers for the new economy: in New York, greengrocers are now mainly Korean; the plumbers fitting gentrified buildings are often Italian, the carpenters Polish; the domestic workers and nannies looking after the houses and children of gentrifiers come from El Salvador, Barbados or elsewhere in the Caribbean.

Immigrants come to the city from every country where US capital has opened markets, disrupted local economies, extracted resources, removed people from the land, or sent the marines as a "peace-keeping force" (Sassen 1988). This global dislocation comes home to roost in the "Third-Worlding" of the US city (Franco 1985; Koptiuch 1991), which, combined with the threat of increasing crime and repressive policing of the streets, invites visions of a predaceous assault on the very gentrification that it helped to stimulate. In her research on the disruption of the ways in which children are socialized, Cindi Katz (1991a, 1991b) finds a clear parallel between the streets of New York and the fields of Sudan where an agricultural project has come to town. The "primitive" conditions of the core are at once exported to the periphery while those of the periphery are reestablished at the core. "As if straight out of some sci-fi plot," writes Koptiuch (1991), "the wild frontiers dramatized in early travel accounts have been moved so far out and away that, to our unprepared astonishment, they have imploded right back in our midst." It is not just the Indian wars of the Old West that have come home to the cities of the East, but the new global wars of the New American World Order.

A new social geography of the city is being born but it would be foolish to expect that it will be a peaceful process. The attempt to reclaim Washington, DC (probably the most segregated city in the US), through white gentrification is widely known by the African-American majority as "the Plan." In London's gentrifying Docklands and East End, an anarchist gang of unemployed working-class kids justify mugging as their "yuppie tax," giving a British twist to the Tompkins Square slogan, "Mug a yuppie." As homes and communities are converted into a new frontier, there is an often clear perception of what is coming as the wagons are circled around. Frontier violence comes with cavalry charges down city streets, rising official crime rates, police racism and assaults on the "natives." And it comes with the periodic torching of homeless people as they sleep, presumably to get them "out of sight." And it comes with the murder of Bruce Bailey, a Manhattan tenant activist, in 1989: his dismembered body was found in garbage bags in the Bronx, and, although police openly suspected angry landlords of the crime, no one was ever charged. It is difficult to be optimistic that the next wave of gentrification will bring a new urban order more civilized than the first.

IS GENTRIFICATION A DIRTY WORD?

On the morning of December 23, 1985, *New York Times* readers awoke to find the most prestigious advertising spot in their morning paper taken up by an editorial advert in praise of gentrification. Some years earlier the newspaper had begun to sell the bottom right quarter of its Opinion Page to the Mobil Corporation, which used it to extol the social and cultural merits of organized global capitalism. By the mid-1980s, with the New York real estate market ablaze, gentrification was increasingly understood as a threat to people's rents, housing and communities, and the Mobil Corporation no longer had an exclusive claim to the purchased ideological ink of the *Times'* Opinion Page. It was "The Real Estate Board of New York, Inc." which now purchased the space in order to bring a defense of gentrification to the citizens of New York. "There are few words in a New Yorker's vocabulary that are as emotionally loaded as 'gentrification,'" the advert began. Gentrification means different things to different people, the Real Estate Board conceded, but "In simple terms, gentrification is the upgrading of housing and retail businesses in a neighborhood with an influx generally of *private* investment." It is a contributor to the diversity, the great mosaic of the city, the advert suggested; "neighborhoods and lives blossom." If a modicum of displacement inevitably results from a neighborhood's private market "rehabilitation," suggests the Board, "We believe" that it "must be dealt with with public policies that promote low- and moderate-income housing construction and rehabilitation, and in zoning revisions that permit retail uses in less expensive, side street locations." It concludes: "We also believe that New York's best hope lies with families, businesses and lending institutions willing to commit themselves for the long haul to neighborhoods that need them. That's gentrification." This was an astonishing declaration, not so much for the predictable ideological tones of what it said but for the fact that it was said at all. How did it come about that the very powerful Real Estate Board of New York, Inc. – the professional lobby for the city's largest real estate developers, a kind of chamber of commerce for promoting real estate interests – found itself in such a defensive position that it had to take out an advertisement in the *Times* for the purpose of trying to redefine one of its major preoccupations? How had gentrification become such a contested issue that its proponents had to summon the full ideological complement of "family" and the private market in its defense?

IS GENTRIFICATION A DIRTY WORD?

There are few words in a New Yorker's vocabulary that are as emotionally loaded as "gentrification."

To one person, it means improved housing. To another, it means unaffordable housing. It means safer streets and new retail businesses to some. To others, it means the homogenization of a formerly diverse neighborhood. It's the result of one family's drive for home ownership. It's the perceived threat of higher rental costs for another family.

In simple terms, gentrification is the upgrading of housing and retail businesses in a neighborhood with an influx of private investment. This process and its consequences, however, are rarely simple.

Neighborhoods and lives blossom.

Examples of gentrification are as varied and distinctive as New York itself and reflect the city's enduring vitality. That vitality is expressed in terms of change...for neighborhoods and people. We see immigrants from Asia transforming the Flushing community in Queens with their industriousness, while recent arrivals from Russia are bringing new flavor to the Brighton Beach area of Brooklyn. Over a decade ago, painters, sculptors and fledgling dance companies looking for loft space turned SoHo, then a manufacturing "ghost town" on Lower Manhattan's northern border, into a world-renowned artistic center. Today a new generation of artists is creating a similar colony in Greenpoint, Brooklyn. Elsewhere, middle class pioneers have bought brownstones in dilapidated areas and enlivened their districts—such as the portion of Columbus Avenue north of Lincoln Center for the Performing Arts—with energy and style.

Different neighborhoods throughout the city have undergone similar changes at different times: Park Slope, Chelsea and the Upper West Side, for example. In each case, neighborhoods that were under-populated and had become shabby and/or dangerous were turned into desirable addresses by families and merchants willing to risk their savings and futures there.

Who has to make room for gentrification?

The greatest fears inspired by gentrification, of course, are that low-income residents and low-margin retailers will be displaced by more affluent residents and more profitable businesses.

The Department of City Planning's study of gentrified neighborhoods in Park Slope and on the Upper West Side concluded that some displacement occurs following a community's decline as well as after its rehabilitation. The study also found, however, that residential rent regulations gave apartment dwellers substantial protection against displacement. In addition, the study pointed out that the mix of retail stores and service establishments has remained the same in both areas since 1970.

In this regard, it should also be noted that tenants of residential rental buildings that are converted to cooperative ownership remain protected by non-eviction plans if they decide they don't want to buy their units. A survey conducted by the Real Estate Board of New York found that 85 percent of such tenants thought the conversion process had been a fair one.

A role for public policy.

We believe that whatever displacement gentrification causes, though, must be dealt with with public policies that promote low- and moderate-income housing construction and rehabilitation, and in zoning revisions that permit retail uses in less expensive, side street locations.

We also believe that New York's best hope lies with families, businesses and lending institutions willing to commit themselves for the long haul to neighborhoods that need them.

That's gentrification.

The Real Estate Board of New York, Inc.

Plate 2.1 "Is Gentrification a Dirty Word?" (Real Estate Board of New York, Inc.)

As I read this ad, propped up in bed, I reflected on how much things had changed in barely ten years. As an undergraduate, visiting the US from small-town Scotland, I began doing gentrification research in 1976 in Philadelphia. In those days I had to explain to everyone – friends, fellow students, professors, casual acquaintances, smalltalkers at parties – what precisely this arcane academic term meant. Gentrification is the process, I would begin, by which poor and working-class neighborhoods in the inner city are refurbished via an influx of private capital and middle-class homebuyers and renters – neighborhoods that had previously experienced disinvestment and a middle-class exodus. The poorest working-class neighborhoods are getting a remake; capital and the gentry are coming home, and for some in their wake it is not entirely a pretty sight. Often as not that ended the conversation, but it also occasionally led to exclamations that gentrification sounded like a great idea: had I come up with it?

Less than ten years later gentrification's notoriety had caught up with the process itself, a process that was well under way in many cities since the late 1950s and early 1960s. From Sydney to Hamburg, Toronto to Tokyo, activists, tenants, everyday people now knew exactly what gentrification was and how it affected their daily lives. It was increasingly recognized for what it was: a dramatic yet unpredicted reversal of what most twentieth-century urban theories had been predicting as the fate of the central and inner city. As such the process was so publicly contested, in the pages of newspapers, popular magazines, academic journals and in the streets, that in the middle of the most intense wave of gentrification to affect the city, the most prestigious advertising space in the *New York Times* was purchased by the city's developers, who felt obliged to defend their gentrification of the city.

The language of gentrification proved irresistible. For those broadly opposed to the process and its deleterious effect on poor residents in affected areas, or even those who were simply suspicious, this new word, gentrification, captured precisely the class dimensions of the transformations that were under way in the social geography of many central and inner cities. Many of those who were more sympathetic to the process resorted to more anodyne terminology – "neighborhood recycling," "upgrading," "renaissance," and the like – as a means to blunt the class and also racial connotations of "gentrification," but many were also attracted by the seeming optimism of "gentrification," the sense of modernization, renewal, an urban cleansing by the white middle classes. The postwar period, after all, had intensified the rhetoric of disinvestment, dilapidation, decay, blight and "social pathology" applied to central cities throughout the advanced capitalist world. If this "discourse of decline" (Beauregard 1993) was most acute in the US, as perhaps befitted the experience of decline and ghettoization, it nevertheless had a broad applicability and invocation.

The language of revitalization, recycling, upgrading and renaissance suggests that affected neighborhoods were somehow devitalized or culturally moribund prior to gentrification. While this is sometimes the case, it is often also true that very vital working-class communities are culturally *de*vitalized through gentrification as the new middle class scorns the streets in favor of the dining room and bedroom. The idea of "urban pioneers" is as insulting applied to contemporary cities as the original idea of "pioneers" in the US West. Now, as then, it implies that no one lives in the areas being pioneered – no one worthy of notice, at least. In Australia the process is known as trendification, and elsewhere, inmovers are referred to as the "hipeoisie." The term gentrification expresses the obvious class character of the process, and for that reason, although it may not be technically a "gentry" that move in but rather middle-class white professionals, it is most realistic.

As is now well documented, "gentrification" was coined by the eminent sociologist, Ruth Glass, in London in 1964. Here is her classic definition and description:

> One by one, many of the working-class quarters of London have been invaded by the middle classes – upper and lower. Shabby, modest mews and cottages – two rooms up and two down – have been taken over, when their leases have expired, and have become elegant, expensive residences. Larger Victorian houses, downgraded in an earlier or recent period – which were used as lodging houses or were otherwise in multiple occupation – have been upgraded once again. . . . Once this process of "gentrification" starts in a district it goes on rapidly until all or most of the original working-class occupiers are displaced and the whole social character of the district is changed.
>
> (Glass 1964: xviii)

The critical intent of Glass's coinage is unmistakable, and was widely understood as the word passed into common usage. It was precisely this critical intent that developers, landlords and the Real Estate Board had been unable to blunt, despite the vigorous promotion of more neutral-sounding euphemisms to script the class and race contours of gentrification. With its 1985 ad, the Real Estate Board, having failed to sink the word, now sought to redefine it, give it a new, less emotional charge, gentrifying the word itself. And they were not alone. At the ground breaking for a major gentrification project in Harlem, only two months prior to the Real Estate Board's advert, New York Senator Alfonse D'Amato, an exuberant defender and benefactor of real estate capital, responded angrily to demonstrators that gentrification equalled nothing more nor less than "housing for working people."

The cachet of "gentrification," however, has been too great for the word and its meanings not to travel – sometimes in astonishing ways. For example, in a newspaper report on new paleontological evidence about the advance of domesticated agriculture into Europe some 9,000 years ago, at the expense of hunter-gatherers, the following account is given, including a quote from a British academic: "The hunter-gatherers who stood in the path of the advance 'suffered a process of gentrification – or even yuppification – from the east'" (Stevens 1991). Less a stretch of the imagination, perhaps, is the following critical collapse of all new history into the experience of New York's gentrifying "East Village":

> When "history" overtakes some new chunk of the recent past, it always comes as a relief – one thing that history does . . . is to fumigate

experience, making it safe and sterile. . . . Experience undergoes eternal gentrification; the past, all the parts of it that are dirty and exciting and dangerous and uncomfortable and real, turns gradually into the East Village.

("Notes and comment" 1984: 39; see also Lowenthal 1986: xxv)

The symbolic power of "gentrification" means that this kind of generalization of meaning is surely inevitable, but even when it takes place in a critical vein, this is a mixed blessing. As with all metaphors, "gentrification" can be used to impart a critical (or not so critical) inflection on radically different experiences and events. But "gentrification" itself is in turn inflected by its metaphorical appropriation: to the extent that "gentrification" is generalized to stand for the "eternal" inevitability of modern renewal, the renovation of the past, the sharply contested class and race politics of contemporary gentrification are dulled. Opposition to gentrification here and now can too quickly be dismissed as a hunter-gatherer rejection of "progress." In fact, for those impoverished, evicted or made homeless in its wake, gentrification is indeed a dirty word and it should stay a dirty word.

A SHORT HISTORY OF GENTRIFICATION

Although the emergence of gentrification proper can be traced to the postwar cities of the advanced capitalist world, there are significant precursors. In his well-known poem, "The Eyes of the Poor," Charles Baudelaire wraps a proto-gentrification narrative into a poem of love and estrangement. Set in the late 1850s and early 1860s, amid Baron Haussmann's destruction of working-class Paris and its monumental rebuilding (see Pinkney 1972), the poem's narrator tries to explain to his lover why he feels so estranged from her. He recalls a recent incident when they sat outside a "dazzling" cafe, brightly lit outside by gaslight, making its debut. The interior was less alluring, decorated with the ostentatious kitsch of the day: hounds and falcons, "nymphs and god-desses bearing piles of fruits, pâtés and game on their heads," an extravagance of "all history and all mythology pandering to gluttony." The cafe stood at the corner of a new boulevard which was still strewn with rubble, and as the lovers swoon in each other's eyes, a bedraggled poor family – father, son and baby – stops in front of them and stares large-eyed at the spectacle of consumption. "How beautiful it is!" the son seems to be saying, although no words were uttered: "But it is a house where only people who are not like us can go." The narrator feels "a little ashamed of our glasses and decanters, too big for our thirst," and for a moment connects in empathy with "the eyes of the poor." Then he turns back to his lover's eyes, "dear love, to read *my* thoughts there." But instead he sees only disgust in her eyes. She bursts out: "Those people with their great saucer eyes are unbearable! Can't you go tell the manager to get them away from here?" (Baudelaire 1947 edn. no. 26).

Marshall Berman (1982: 148–150) uses this poem to introduce his discussion of "modernism in the streets," equating this early *embourgeoisement* of Paris (Gaillard 1977; see also Harvey 1985a) with the rise of bourgeois modernity. Much the same connection was made at the time, albeit across the English Channel. Eighty years before Robert Park and E. Burgess (Park *et al.* 1925) developed their influential "concentric ring" model for the urban structure of Chicago, Friedrich Engels made a similar generalization concerning Manchester:

> Manchester contains, at its heart, a rather extended commercial district, perhaps half a mile long and about as broad, and consisting almost completely of offices and warehouses. Nearly the whole district is abandoned by dwellers. . . . This district is cut through by certain main thoroughfares upon which the vast traffic concentrates, and in which the ground level is lined with brilliant shops. . . . With the exception of this commercial district, all Manchester proper [comprises] unmixed working-people's quarters, stretching like a girdle, averaging a mile and a half in breadth, around the commercial district. Outside, beyond this girdle, lives the upper and middle bourgeoisie.
>
> (Engels 1975 edn.: 84–85)

Engels had a keen sense of the social effects of this urban geography, especially the efficient concealment of "grime and misery" from "the eyes of the wealthy men and women" residing in the outer ring. But he also witnessed the so-called "Improvements" of mid-nineteenth-century Britain, a process for which he chose the term "Haussmann." "By the term 'Haussmann,'" he explained, "I do not mean merely the specifically Bonapartist manner of the Parisian Haussmann" – the Prefect of Paris, who was building boulevards through the "closely built workers' quarters and lining them on both sides with big luxurious buildings," for the strategic purpose of "making barricade fighting more difficult," and for turning "the city into a luxury city pure and simple" (Engels 1975 edn.: 71). Rather, he suggested, this was a more general process:

> By "Haussmann" I mean the practice, which has now become general, of making breaches in the working-class quarters of our big cities, particularly in those which are centrally situated, irrespective of whether this practice is occasioned by considerations of public health and beautification or by demand for big, centrally located business premises or by traffic requirements. . . . No matter how different the reasons may be, the result is everywhere the same: the most scandalous alleys and lanes disappear to the accompaniment of lavish self-glorification by the bourgeoisie on account of this tremendous success.
>
> (Engels 1975 edn.: 71)

Earlier examples of gentrification have been cited. Roman Cybriwsky, for example, provides a nineteenth-century print depicting a family's displacement from a tenement in Nantes in 1685. He reports that the Edict of Nantes, signed by Henry IV in 1598, guaranteed poor Huguenots certain rights including access to housing, but when the edict was revoked nearly a century later by Louis XIV, wholesale displacement took place at the hands of landlords, merchants and wealthier citizens (Cybriwsky 1980). Be that as it may, something more akin to contemporary gentrification made an appearance in the middle of the nineteenth century, whether known by the name "*embourgeoise-ment*," "Haussmann" or the "Improvements." It was hardly "general," to use Engels' word, but sporadic, and it was surely restricted to Europe since few cities in North America, Australia or elsewhere had the extent of urban history to provide whole neighborhoods of disinvested stock. Chicago was barely ten years old when Engels made his first observations of Manchester; and as late as 1870, there was little urban development in Australia. The closest parallel in North America might be the process whereby one generation of wooden buildings was quickly torn down to be replaced by brick structures and these in turn – at least in the older east-coast cities – were demolished to make room for larger tenements or single-family houses. It would be misleading to consider this gentrification, however, insofar as such redevelopment was an integral part of the outward geographical expansion of the city and not, as with gentrification, a spatial reconcentration.

Even as late as the 1930s and 1940s, gentrification remained a sporadic occurrence, but by this time precursor experiences of gentrification were also turning up in the United States. The flavor remained resolutely European and

Plate 2.2 "Persecution after the Edict of Nantes": print by Jules Girardet, 1885
(courtesy of Roman Cybriwsky)

aristocratic, however, laced through with liberal guilt. The spirit of the enterprise is well captured in a recent retrospective by Maureen Dowd, recalling the Georgetown scene in Washington, DC's most gentrified neighborhood through the eyes of patrician hostess turned historian Susan Mary Alsop:

> They gentrified Georgetown, an unfashionable working-class neighborhood with a large black contingent. As Mrs. Alsop told *Town and Country* magazine: "The blacks kept their houses so well. All of us had terrible guilt in the 30's and 40's for buying places so cheaply and moving them out."
>
> The gentry and the hostesses faded through the 1970s.
>
> (Dowd 1993: 46)

Similar scenes were being lived out in Boston's Beacon Hill (Firey 1945), albeit with a different local flavor, or for that matter in London, although of course genteel society had in no way relinquished its claim to many London neighborhoods in quite the same fashion.

So what makes all of these experiences "precursors" to a gentrification process that began in earnest in the postwar period? The answer lies in both the extent and the systemic nature of central and inner-city rebuilding and rehabilitation beginning in the 1950s. The nineteenth-century experiences in London and Paris were unique, resulting from the confluence of a class politics aimed at the threatening working classes and designed to consolidate bourgeois control of the city, and a cyclical economic opportunity to profit from rebuilding. The "Improvements" were certainly replicated in different ways and at a lesser scale in some other cities – Edinburgh, Berlin, Madrid, for example – but, as in London and Paris, they were historically discrete events. There are no systematic "improvements" in London in the first decades of the twentieth century, or a continued *embourgeoisement* of Paris in the same period systematically altering the urban landscape. As regards the incidences of gentrification in the mid–twentieth century, these were so sporadic that the process was unknown in the majority of large cities. It was very much an exception to larger urban geographic processes. Its agents, as in the case of Georgetown or Beacon Hill, were generally from such a limited social stratum and in many cases so wealthy that they could afford to thumb their patrician noses at the mere dictates of the urban land market – or at least mold the local market to their wonts.

This all begins to change in the postwar period, and it is no accident that the word "gentrification" is coined in the early 1960s. In Greenwich Village in New York, where gentrification was associated with a nascent counterculture; in Glebe in Sydney, where sustained disinvestment, rental deregulation, an influx of southern European immigrants, and the emergence of a middle-class resident action group all conspired toward gentrification (B. Engels 1989); in Islington in London where the process was relatively decentralized; and in dozens of other large cities in North America, Europe and Australia, gentrification began to occur. And nor was this process long confined simply to the largest cities. By 1976, one study concluded that nearly half of the 260 US cities with a population of more than 50,000 were experiencing gentrification (Urban Land Institute 1976). Barely twelve years after Ruth Glass had coined the term, it was no longer just New York, London and Paris that were being gentrified, but Brisbane and Dundee, Bremen and Lancaster, PA.

Gentrification today is ubiquitous in the central and inner cities of the advanced capitalist world. As unlikely a city as Glasgow, simultaneously a symbol and stronghold of working-class grit and politics, was sufficiently gentrified by 1990, in a process fueled by an aggressive local state, to be adopted as "European City of Culture" (Jack 1984; Boyle 1992). Pittsburgh and Hoboken are perhaps US equivalents. In Tokyo, the central ward of Shinjuku, once a meeting place for artists and intellectuals, has become a "classic battleground" of gentrification amid a rampaging real estate market (Ranard 1991). Likewise Montparnasse in Paris. Prague's response to an unleashed real estate market since 1989 has been torrid gentrification, almost on the scale of Budapest's (Sýkora 1993), while in Madrid it was the end of Franco's fascism and a comparative democratization of urban government that cleared the way for reinvestment (Vázquez 1992). In the Christianhavn area around the experimental "free city" of Christiania on Copenhagen's waterfront (Nitten 1992), and in the back streets of Granada adjacent to the Alhambra, gentrification proceeds in tense affinity with tourism. Even outside the most developed continents – North America, Europe and Australasia – the process has begun to take place. In Johannesburg, the gentrification of the 1980s (Steinberg et al. 1992) has been significantly attenuated by a new kind of "white flight" since the election of the ANC in April 1994 (Murray 1994: 44–48), but the process has also affected smaller cities such as Stellenbosch (Swart 1987). In São Paulo a very different pattern of disinvestment in land has taken place (Castillo 1993), but a modest renovation and reinvestment in the Tatuapé district accommodates small business owners and professionals who work in the central business district but who can no longer afford the rapidly inflating prices of the most prestigious central enclaves such as Jardin. Much of this redevelopment involves "verticalization" (Aparecida 1994) as land served by basic services is scarce. More generally, the "middle zones" around São Paulo and Rio de Janeiro are experiencing development and redevelopment for the middle class (Queiroz and Correa 1995: 377–379).

Not only has gentrification become a widespread experience since the 1960s, then, but it is also systematically integrated into wider urban and global processes, and this too differentiates it from earlier, more discrete experiences of "spot rehabilitation." If the process that Ruth Glass observed in London at the beginning of the 1960s, or even the planned remake of Philadelphia's Society Hill during the same period, represented somewhat isolated developments in the land and housing markets, they did not remain so. By the 1970s gentrification was clearly becoming an integral residential thread in a much larger urban restructuring. As many urban economies in the advanced capitalist world experienced the dramatic loss of manufacturing jobs and a parallel increase in producer services, professional employment and the expansion of so-called "FIRE" employment (Finance, Insurance, Real Estate), their whole urban geography underwent a concomitant restructuring. Condominium and cooperative conversions in the US, tenure conversions in London and international capital investments in central-city luxury accommodations were increasingly the residential component of a larger set of shifts that brought an office boom to London's Canary Wharf (A. Smith 1989) and New York's Battery Park City (Fainstein 1994) and the construction of new recreational and retail landscapes from Sydney's Darling Harbour to Oslo's AckerBrygge. These economic shifts were often accompanied by political shifts as cities found themselves competing in the global market, shorn of much of the traditional protection of national state institutions and regulations: deregulation, privatization of housing and urban services, the dismantling of welfare services – in short, the remarketization of public functions – quickly followed, even in bastions of social democracy such as Sweden. In this context, gentrification became a hallmark of the emerging "global city" (Sassen 1991), but was equally a presence in national and regional centers that were themselves experiencing an economic, political and geographical restructuring (M. P. Smith 1984; Castells 1985; Beauregard 1989).

In this regard, what we think of as gentrification has itself undergone a vital transition. If in the early 1960s it made sense to think of gentrification very much in the quaint and specialized language of residential rehabilitation that Ruth Glass employed, this is no longer so today. In my own research I began by making a strict distinction between gentrification (which involved rehabilitation of existing stock) and redevelopment that involved wholly new construction (N. Smith 1979a), and at a time when gentrification was distinguishing itself from large-scale urban renewal this made some sense. But I no longer feel that it is such a useful distinction. Indeed 1979 was already a bit late for this distinction. How, in the larger context of changing social geographies, are we to distinguish adequately between the rehabilitation of nineteenth-century housing, the construction of new condominium towers, the opening of festival markets to attract local and not so local tourists, the proliferation of wine bars – and boutiques for everything – and the construction of modern and postmodern office buildings employing thousands of professionals, all looking for a place to live (see, for example, A. Smith, 1989)? This after all describes the new landscapes of downtown Baltimore or central Edinburgh, waterfront Sydney or riverside Minneapolis. Gentrification is no longer about a narrow and quixotic oddity in the housing market but has become the leading residential edge of a much larger endeavor: the class remake of the central urban landscape. It would be anachronistic now to exclude redevelopment from the rubric of gentrification, to assume that the gentrification of the city was restricted to the recovery of an elegant history in the quaint mews and alleys of old cities, rather than bound up with a larger restructuring (Smith and Williams 1986).

Having stressed the ubiquity of gentrification at the end of the twentieth century, and its direct connection to fundamental processes of urban economic, political and geographical restructuring, I think it is important to temper this vista with a sense of context. It would be foolish to think that the partial geographical reversal in the focus of urban reinvestment implies the converse, the end of the suburbs. Suburbanization and gentrification are certainly interconnected. The dramatic suburbanization of the urban landscape in the last century or more provided an alternative geographical locus for capital accumulation and thereby encouraged a comparative disinvestment at the center – most intensely so in the US. But there is really no sign that the rise of gentrification has diminished contemporary suburbanization. Quite the opposite. The same forces of urban restructuring that have ushered new landscapes of gentrification to the central city have also transformed the suburbs. The recentralization of office, retail, recreation and hotel functions has been accompanied by a parallel decentralization which has led to much more functionally integrated suburbs with their own more or less urban centres – edge cities as they have been called (Garreau 1991). If suburban development has in most places been more volatile since the 1970s in response to the cycles of economic expansion and contraction, suburbanization still represents a more powerful force than gentrification in the geographical fashioning of the metropolis.

From the 1960s to the 1990s, however, as academic and political critiques of suburbanization were mounting, gentrification for many came to express an extraordinary optimism, warranted or otherwise, concerning the future of the city. The urban uprisings and social movements of the 1960s notwithstanding, gentrification represented a wholly unpredicted novelty in the urban landscape, a new set of urban processes that took on immediate symbolic importance. The contest over gentrification represented a struggle not just for new and old urban spaces but for the symbolic political power to determine the urban future. The contest was as intense in the newspapers as it was in the streets, and for every defense of gentrification such as that by the Real Estate Board of New York there was an assault against gentrification-induced displacement, rent increases and neighborhood change (see, for example, Barry and Derevlany 1987). But the contest over gentrification was also played out in the usually more bromidic pages of academic journals and books.

THE GENTRIFICATION DEBATES: THEORIES OF GENTRIFICATION OR THE GENTRIFICATION OF THEORY?

The emergence of gentrification has unleashed a remarkably lively debate in scholarly circles since the early 1980s. Chris Hamnett has argued that there are several key reasons for this. In the first place, as we have already argued, gentrification represents a novel set of processes and "one of the major 'leading edges' of contemporary metropolitan restructuring" (Hamnett 1991: 174). Second, insofar as gentrification results in significant displacement, this has raised questions concerning appropriate urban policy. Third, gentrification clearly challenges traditional theories from the Chicago School, social ecology tradition or postwar positivist school of urban economics (see for example Alonso 1964). In none of these traditions could a "return to the city" be adequately foreseen. Finally, gentrification became "a key theoretical and ideological battleground" between those stressing culture and individual choice, consumption and consumer demand on the one side and others emphasizing the importance of capital, class and the impetus of shifts in the structure of social production (Hamnett 1991: 173–174).

These debates have consumed a lot of time and ink, but, as the above multiplicity of ingredients suggests, the stakes were considerable. Hamnett is surely correct to argue that the final reason – a recognition that gentrification constituted an intense ideological as well as theoretical battlefield – may be the crucial one. Much of the early research on gentrification, especially in the US, involved case studies that largely adhered to the implicit assumptions of post-war urban theory (Lipton 1977; Laska and Spain 1980). In particular, they adopted what came to be referred to as a "consumption-side" explanation whereby the gamut of "neighborhood change" is to be explained primarily in terms of who moves in and who moves out. Alternative explanations emphasized the role of the state (Hamnett 1973) in encouraging gentrification and the importance of financial institutions in selectively providing the capital for rebuilding (Williams 1976, 1978). This production-side explanation received further impetus from a consideration of capital *dis*investment, the role of disinvestment in establishing the opportunity for gentrification, the proposal of a "rent gap" theory, and by the location of gentrification within a broader theoretical perspective of "uneven development" (N. Smith 1979a, 1982). At the same time, the simplicity of the consumption-side argument was being superseded by efforts to see consumption too in the wider context of middle-class ideology and "post-industrial society" (Ley 1978, 1980).

Doubled up on this theoretical contest was a political one. Consumption-side arguments were at times presented by quite conservative voices in the urban literature, although many conservatives also simply dismissed gentrification as a momentary and insignificant process (Berry 1985; Sternlieb and Hughes 1983). More often, the consumption-side position was adopted by political liberals who broadly celebrated the advent of a postindustrial city and the rehabilitation of slum neighborhoods while lamenting the social costs. Insofar as they focused on class it was the middle class, often a new middle class, who were vaunted as the subjects of history. By contrast, production-side explanations were more usually advanced by adherents of radical social theory, including marxism, for whom gentrification was symptomatic of a wider class geography of the city which was continually replicated and reinvented in various ways, including the patterns and rhythms of capital investment in housing.

A flurry of debates ensued in the 1980s and continued into the 1990s, pitting production-side against consumption-side explanations; proposing the cultural rather than capitalist roots of gentrification; exploring the importance of the changed social position of women for an explanation of gentrification; identifying rent gaps (see Chapter 3); rejecting, reconsidering and restating the rent gap theory; explaining the "gentrifiers"; critiquing ideologies of gentrification, and so forth. This is not the place to review what became a vibrant, complicated, sometimes counterproductive set of arguments, claims and counterclaims.[1] As one of the participants I have very definite opinions on

Plate 2.3 The Art of Eviction: an art exhibit at ABC No Rio in the Lower East Side, New York

the course of the debate, but it seems to me that by the mid-1980s there was indeed some reconciliation between these opposing explanations. It has been my contention – and I think the claim of many others, including David Ley – that explanations which remained confined to consumption or production practices, narrowly conceived, were of decreasing relevance (Smith and Williams 1986, Ley 1986). Likewise, the integration of cultural and capital-centered explanations is vital, in precisely the manner pioneered by Sharon Zukin (1982, 1987). These and other calls (Lees 1994) for a more integrative approach have been aided by Clark's (1991, 1994) appeals for a recognition of the complementarity of differing explanations. And yet many of the original theoretical and political fault lines remain, albeit in altered states. Indeed they have been reasserted via the recent fashionability of postmodernism in urban theory.

If Baudelaire, Engels and Berman (1982) all saw the Haussmannization of

Paris as one defining moment of a capitalist modernity, can we see in gentrification a defining geography of postmodernity? There is considerable debate about the extent to which it is accurate or even useful to describe the urban and economic restructurings of the post-1970s era as a shift from Fordist to post-Fordist forms of political economic regulation, from a more rigid to a more flexible mode of accumulation (see, for example, Gertler 1988; Reid 1995). Likewise the connection between this flexibility in the economic sphere and the emergence of postmodernism in cultural terms is also subject to debate (Harvey 1989). Although they would not wish to connect their cultural concerns to the economic argument in this way, some theorists have in different ways proposed that we ought to conceive of gentrification as a postmodern urbanism (Mills 1988, Caulfield 1994; Boyle 1995). For these authors it is less a question of developing the connections, inherent in gentrification, between economic and cultural shifts, resulting in a new urban geography. Rather, in this vision culture virtually supplants economics, and agency can be distilled down to the narrowest philosophical individualism (Hamnett 1992). Gentrification is reconfigured as an expression of personal activism by the new middle classes, their personal triumph of culture over economics. Only via such a cultural determinism and via a wholesale and uncritical alignment with the subject position of the "gentrifiers" can one celebrate gentrification, as Caulfield (1989: 628) does, as constituting "emancipatory practice" and expressing "the space of freedom and critical spirit of the city."

This is surely Foucault run amok: "If it moves it must be political, and emancipatory to boot." If gentrification is emancipatory political practice, it is difficult to see it as anything other than political activism *against* the working class. This extreme proposal of postmodern urbanism is surely less a contribution to theories of gentrification than to the gentrification of theory.

The postmodern and poststructuralist concern with subject positionality began as a very useful and necessary means to "decenter" the universal subject in social, political and cultural discourse. In some treatments, however, the postmodern turn has come full circle. There is no doubt that postmodernism has helped to incubate a serious analysis of the cultural dimensions of urban change that had hitherto been lacking. In the appropriation of postmodernism as a script for gentrification, "postmodernism urbanism" has for many passed into a vehicle for the radical *re*centering of the subject on the author him- or herself.[2] If decentering taught us that the author was in the world rather than somehow above it, and encouraged us to see the world in the author, a rather reactionary version of postmodernism flips the equation: "we *are* the world." Bob Fitch has put it best:

> Under the influence of the postmodern mood, the left has generated a new political grammar. The political subject has changed. It is no longer the masses, workers, the people. Them. Nowadays it's us. It is the left intelligentsia itself which has become the subject of political activity. Our concerns, not theirs.

> (Fitch 1988: 19)

Far from opposing the evictions, rent gouging, displacement, homelessness, violence and other class-exploitative and class-abusive practices that gentrification brings, more extreme proclamations of a postmodern urbanism simply gentrify the working class out of the picture. We, the middle-class authors, recognizing that our own "activism" has become so digressive, desperately reinvent that activism as the magic explanation *and justification* for gentrification itself. Agency is safely restored to the middle class – laced through with emancipatory piety – and the working class are disappeared.

I hope it will be clear enough from the essays in this book that my political barbs here are aimed at a particularly opportunistic version of postmodernism and not at the so-called cultural turn *per se*. Cultural analysis is vital to the explanation of gentrification, but there are different kinds of cultural analyses (Mitchell 1995b). Cultural analyses also occur "in the world," and the luxury of omitting the violence of gentrification from our cultural purview is a political luxury born of class and race privilege.

In 1969, the sociologist Martin Nicolaus made a novel proposal that I have always felt to be inspired. Combating the objectivist and controlling gaze of mainstream 1960s sociology, Nicolaus proposed an alternative vision for sociocultural research – a vision which made explicit rather than implicit the social position from whence it came:

> What if that machinery were reversed? What if the habits, problems, actions, and decisions of the wealthy and powerful were daily scrutinized by a thousand systematic researchers, were hourly pried into, analyzed, and cross-referenced, tabulated and published in a hundred inexpensive mass-circulation journals and written so that even the fifteen-year-old high school drop-outs could understand it and predict the actions of their parents' landlord, manipulate and control him?

It seems to me that to the extent that gentrification research focuses on the so-called gentrifiers themselves – only ever a part of the equation – the discernment of this vision provides an excellent starting point.

THE REVANCHIST CITY

As the 1980s drew to a close and President George Bush was promising the US public a "kinder, gentler nation," US cities were headed in a diametrically opposite direction. If gentrification had spearheaded a certain middle-class optimism about the city, the end of the 1980s boom, the crystallized effects of a decade of deregulation, privatization and emerging cuts in welfare and social service budgets rewrote the urban future as one of gloom, not boom (Fitch 1993). As if this were not enough, severe economic crisis and governmental retraction were emulsified by a visceral reaction in the public discourse against the liberalism of the post-1960s period and an all-out attack on the social policy structure that emanated from the New Deal and the immediate postwar era. Revenge against minorities, the working class, women, environmental legislation, gays and lesbians, immigrants became the increasingly common denominator of public discourse. Attacks on affirmative action and immigration policy, street violence against gays and homeless people, feminist bashing and public campaigns against political correctness and multi-culturalism were the most visible vehicles of this reaction. In short, the 1990s have witnessed the emergence of what we can think of as the *revanchist city* (N. Smith 1996a).

Revanche in French means revenge, and the revanchists comprised a political movement that formed in France in the last three decades of the nineteenth century. Angered by the increased liberalism of the Second Republic, the ignominious defeat to Bismarck, and the last straw – the Paris Commune (1870–1871), in which the Paris working class vanquished the defeated government of Napoleon III and held the city for months – the revanchists organized a movement of revenge and reaction against both the working class and the discredited royalty. Organized around Paul Déroulède and the Ligue des Patriotes, this movement was as militarist as it was nationalist, but also made a wide appeal to "traditional values." "The True France, for Déroulède – the France of good honest men who believed in simple virtues of honor, family, the army, and the [new Third] Republic . . . would surely win out" (Rutkoff 1981: 23). It was a right-wing movement built on populist nationalism and devoted to a vengeful and reactionary retaking of the country.

The parallels with *fin-de-siècle* France should not be overdrawn, but nor should they be ignored. In the current *fin de siècle* – indeed the *fin de millénaire* – there is a broad, vengeful right-wing reaction against both the "liberalism" of the 1960s and 1970s and the predations of capital. This takes many forms, including fundamentalist religion and a Heideggerian romance of place, precisely at the time when "traditional" identities of place are most threatened by global capital. In the US especially, the public culture and official politics are increasingly an expression of a new creeping revanchism. The Gingrich Congress elected in 1994, the rise of white supremacist militias, the vicious anti-corporatist right-wing populism of Patrick Buchanan, the intense emotion around anti-immigrant campaigns and the call for revenge against beneficiaries of affirmative action all point in this direction.

In many ways the vengefulness of the *fin-de-siècle* revanchist city has overtaken gentrification as a script for the urban future. If, in many places, gentrification was undiminished by the recessions of the early 1980s (Ley 1992), the deeper depression of the late 1980s and early 1990s has severely curtailed gentrification activity in many places, leading many commentators to anticipate *de*gentrification. The bankruptcies of celebrity developers such as Donald Trump in New York or Godfrey Bradman in London and of Olympia and York, the multinational development company which built both Canary Wharf and Battery Park City, confirmed the depth of the real estate crisis of the early 1990s (Fainstein 1994: 61). The language of degentrification emerged first in Manhattan, which experienced a ruthless shakeout of small landlords, developers, marginal real estate agencies, and other gentrification-related businesses between 1989 and 1993. But it was a much more general process. London's Docklands, in the wake of the Canary Wharf bankruptcy, was "left with one of the largest housing lakes in Europe. Unsold developments had to be boarded up for the recession" (McGhie 1994). The result for many people, despite the diminution of gentrification activity, was homelessness, unemployment, enforced squatting, all now to be faced in the context of eviscerated social services.

But it would be a mistake to assume, as the language of degentrification seems to do, that the economic crisis of the early 1990s spelt the secular end of gentrification. Olympia and York and Donald Trump have both restructured – the former entering a salvage partnership with Prince Walid bin Talal of Saudi Arabia, the latter downsizing for an aggressive comeback – and gentrification reemerged on many urban landscapes in the mid-1990s. At best, the depression of the early 1990s brought a reassertion of economics, a more sober set of calculations to the gentrification process than obtained in the 1980s. The language of degentrification can be seen as yet another ploy to redefine or even rid the public discourse of a dirty word, while laying the groundwork for resuming the process that begat it.

And it would also be a mistake to assume that the resumption of gentrification in the late 1990s will militate against the revanchist city. The opposite has become true. As the recent history of Tompkins Square Park and New York's Lower East Side suggests, gentrification has become an integral part of the revanchist city. And if the US in some ways represents the most intense experience of a new urban revanchism, it is a much more widespread experience. Margaret Thatcher prepared the political ground in Britain for an unprecedented decimation of public housing and social services. Gentrification became a major *political* strategy whereby some of London's central boroughs,

Plate 2.4 Bullet Space: an artists' squat in the Lower East Side

such as Westminster and Wandsworth, were widely alleged to have fostered the privatization of public housing in order to move Labour-voting council tenants out and Tory-voting yuppies in. The results were dramatically visible in the May 1990 local elections in England and Wales: "at times it looks as if London is being turned inside out, like a glove. Instead of Tory suburbs and a Labour inner city," suggests one commentator, "Tory voters are reclaiming the city centre and driving the Labour voters out to the fringes" (Linton 1990). Indeed this political reversal was so noticeable that one writer dubbed it the "London effect" (Hamnett 1990).

If eviction of hundreds of squatters in a series of dawn raids in Stamford Hill in April 1991 was much more peaceful than the eviction, in June of that year, of 300 homeless people from New York's Tompkins Square Park, it was not always so. Three years earlier police in Hackney engaged in pitch battles with many of the same squatters. And in Paris, in August 1991, the eviction of squatters from a site adjacent to the partly built national library was accompanied by violent overreaction by the police. Three different attacks in four months in three very different cities, yet a common theme. Or there is Amsterdam, with an even longer and more violent history of squatting and antisquatting attacks. In the struggles that continue in these and many other cities, gentrification and the revanchist city find a common conjuncture in the restructured urban geography of the late capitalist city. The details of each conflict and of each situation may be different, but a broad commonality of contributing processes and conditions set the stage.

The squatters and homeless activists of Paris and London, Amsterdam and New York have made perfectly clear in their own actions that they are fighting a single struggle. If the loss of urban optimism for the middle classes led directly to the new urban revanchism, the resumption of gentrification will further divide and affirm the revanchist city. Persistent warnings of dual or divided cities (Fainstein *et al.* 1992; Mollenkopf and Castells 1991) are surely prescient; that it will simultaneously be a revanchist city makes the new urban frontier a darker and more dangerous prospect. Despite more defeats than victories, there is no sign that squatters and homeless people will suddenly give up the struggle for housing.

Part I

TOWARD A THEORY OF
GENTRIFICATION

LOCAL ARGUMENTS

From "consumer sovereignty" to the rent gap

Following a period of sustained deterioration in the postwar period, many cities began to experience the gentrification of select central and inner-city neighborhoods. Initial signs of revival during the 1950s, most notably in London and New York, intensified in the 1960s, and by the 1970s these had grown into a widespread gentrification movement affecting the majority of the larger and older cities in Europe, North America and Australia. Although gentrification rarely accounts for more than a fraction of new housing starts compared with new construction, the process is very important in those districts and neighborhoods where it occurs. And it has had a very powerful effect on the rethinking of urban cultures and urban futures in the last quarter of the twentieth century.

Gentrification has etched the leading edge of the new urban frontier. If the comprehensive causes and effects of gentrification are rooted in a complex nesting of social, political, economic and cultural shifts, it is my contention here that the complexity of capital mobility in and out of the built environment lies at the core of the process. For all the interpretive cultural optimism that shrouds it, the new urban frontier is also a resolutely economic creation. The causes and effects of gentrification are also complex in terms of scale. While the process is clearly evident at the neighborhood scale it also represents an integral dimension of global restructuring. In this chapter I want to focus on explanations of gentrification at the neighborhood scale; Chapter 4 considers global arguments.

THE LIMITS OF CONSUMER SOVEREIGNTY

As the process of gentrification burgeoned so did the literature about it. The preponderance of this literature concerns the contemporary processes or its effects: the socioeconomic and cultural characteristics, profiles of the new urban immigrants, displacement, the role of the state, benefits to the city, the creation and destruction of community. At least in the beginning, little attempt was made to construct historical explanations of the process, to study causes rather than effects. Instead, explanations were very much taken for granted and have generally fallen into two categories: cultural and economic.

Popular among gentrification theorists is the notion that young, usually professional, middle-class people have changed their lifestyle. According to Gregory Lipton, for example, these changes have been significant enough to "decrease the relative desirability of single-family, suburban homes" (1977: 146). Thus, with the trend toward fewer children, postponed marriages and a fast-rising divorce rate, younger homebuyers and renters are trading in the tarnished dream of their parents for a new dream defined in urban rather than suburban terms. Others have emphasized the search for socially distinctive communities, as in the case of gay gentrification (Winters 1978; Lauria and Knopp 1985), while still others have extended this into a more general argument. In contemporary "post-industrial cities," according to David Ley, where white-collar service occupations supersede blue-collar productive occupations, this brings with it an emphasis on consumption and amenity, not work. Patterns of consumption come to dictate patterns of production; "the values of consumption rather than production guide central city land use decisions" (Ley 1978: 11; 1980). Gentrification is explained as a consequence of this new emphasis on consumption. It represents a new urban geography for a new social regime of consumption. Earlier cultural explanations of this sort have been supplemented more recently by the tendency to treat gentrification as an urban expression of postmodernity or (in more extreme cases) postmodernism (Mills 1988; Caulfield 1994).

Over and against these cultural explanations are a series of closely related economic arguments. As the cost of newly constructed housing has risen rapidly in the postwar city and its distance from the city center increased, the rehabilitation of inner- and central-city structures is seen to be more viable economically. Old properties and housing plots can be purchased and rehabilitated for less than the cost of a comparable new house. In addition, many researchers, especially in the 1970s, stressed the high economic cost of commuting – the higher cost of gasoline for private cars and rising fares on public transportation – and the economic benefits of proximity to work.

These conventional hypotheses are by no means mutually exclusive. They are often invoked jointly and share in one vital respect a common perspective: an emphasis on consumer preference and the constraints within which these preferences are implemented. This assumption of consumer sovereignty is shared with the broader rubric of residential land use theory emanating from postwar neoclassical economics (Alonso 1964; Muth 1969; Mills 1972). According to these theories, suburbanization reflects the preference for space and the increased ability to pay for it due to the reduction of transportational and other constraints. Gentrification, then, is explained as the result of an alteration of preferences and/or a change in the constraints determining which preferences will or can be implemented. Thus in the media and the research literature alike, and especially in the US, where suburbanization bore such a heavy cultural symbolization, gentrification came to be viewed as a "back to the city movement."

This assumption applied as much to the earlier gentrification projects, such as Philadelphia's Society Hill (accomplished after 1959 with substantial state assistance – see Chapter 6), as it does to the later more spontaneous and more ubiquitous emergence of gentrification in the private market (albeit often still with public subsidies). All have become symbolic of a supposed middle- and upper-class pilgrimage back from the suburbs. And yet the pervasive assumption that the gentrifiers are disillusioned suburbanites may not be accurate. As early as 1966, Herbert Gans lamented the lack of any "study of how many suburbanites were actually brought back by urban renewal projects" (1968: 287), and in subsequent years academic studies began to research the issue.

In the first part of this chapter, then, I present some empirical information from Society Hill in Philadelphia as a means of challenging the traditional consumer sovereignty assumptions expressed by the "back-to-the-city" nomenclature. The next section examines the importance of capital investment for the shaping and reshaping of the urban environment, and this is followed by an analysis of disinvestment – a vital but widely ignored determinant of urban change. Finally, I try to bring these themes together in the proposal of a "rent gap" hypothesis for the explanation of gentrification.

A RETURN FROM THE SUBURBS?

The location of William Penn's "holy experiment" in the seventeenth century, Society Hill housed Philadelphia's gentry well into the nineteenth century. With industrialization and urban growth, however, its popularity declined, and the gentry, together with the rising middle class, moved west of Rittenhouse Square, and across the Schuylkill River to West Philadelphia, and to the new suburbs in the northwest. Society Hill deteriorated rapidly toward the end of the nineteenth century, being effectively written off as a "slum" neighborhood (Baltzell 1958). In the 1950s, however, a new city administration aligned itself with a patrician ambition for renewal, and in 1959 an urban renewal plan was implemented. Within ten years Society Hill was transformed. Described seventeen years later in Bicentennial advertising as "the most historic square mile in the nation," Society Hill again came to house the city's middle and upper middle classes and even a few members of the upper classes. Noting the enthusiasm with which rehabilitation was done, the novelist Nathanial Burt captured the elite flavor of many of the early US gentrification projects.

> Remodeling old houses is, after all, one of Old Philadelphia's favorite indoor sports, and to be able to remodel and consciously serve the cause of civic revival all at once has gone to the heads of the upper classes like champagne."
>
> (Burt 1963: 556–557)

As this indoor sport caught on, therefore, it became Philadelphia folklore that "there was an upper class return to center city in Society Hill" (Wold 1975: 325). Burt eloquently explains in the still novel but emerging language of civic boosterism:

The renaissance of Society Hill . . . is just one piece in a gigantic jigsaw puzzle which has stirred Philadelphia from its hundred-year sleep, and promises to transform the city completely. This movement, of which the return to Society Hill is a significant part, is generally known as the Philadelphia Renaissance.

(Burt 1963: 539)

In fact, by June 1962 less than a third of the families purchasing property for rehabilitation were from the suburbs[1] (Greenfield and Co. 1964: 192). But since the first people to rehabilitate houses began work in 1960, it was generally expected that the proportion of suburbanites would rise sharply as the area became better publicized and a Society Hill address became a coveted possession. After 1962, however, no data were officially collected. Table 3.1 presents data sampled from case files held by the Redevelopment Authority of Philadelphia, covering most of the first fifteen years of the project, by which time it was essentially complete. It represents a 17 percent sample of all rehabilitated residences.

It would appear that only a small proportion of gentrifiers – 14 percent – did in fact return from the suburbs to Society Hill. By comparison, 72 percent moved from elsewhere within the city boundaries. A statistical breakdown of this latter group suggests that of previous city dwellers, 37 percent came from Society Hill itself, and 19 percent came from the fashionable Rittenhouse Square district alone. The remainder came largely from several middle- and upper-class neighborhoods in the city: Chestnut Hill, Mount Airy, Spruce Hill. Rather than a return from the suburbs, this would seem to suggest that gentrification is bringing about a recentralization and reconsolidation of upper- and middle-class white residences in the city center. A similar pattern of consolidation can be observed in several of the cities surveyed by Lipton (1977). Additional data from Baltimore and Washington, DC, on the percentage of returning suburbanites support the Society Hill data (Table 3.2). In a European context, similarly, Cortie et al. (1982) find very little evidence of a "return to the city" in connection with the gentrification of the Jordaan district of Amsterdam (see Chapter 8).

Table 3.1 The origin of rehabilitators in Society Hill, Philadelphia, 1964–1975

Year	1964	1965	1966	1969	1972	1975	Total	Percentage by origin
Same address	5	3	1	1	1	0	11	11
Elsewhere in the city	9	17	25	9	12	1	73	72
Suburbs	0	7	4	2	1	0	14	14
Outside SMSA	0	0	0	0	2	0	2	2
Unidentified	0	0	2	0	0	0	2	2
Total	14	27	32	12	16	1	102	100

Source: Redevelopment Authority of Philadelphia case files
Note: SMSA = Standard Metropolitan Statistical Area

Table 3.2 The origin of rehabilitators in three cities

City	Percentage of city dwellers	Percentage of suburbanites
Philadelphia:		
Society Hill	72	14
Baltimore		
Homestead Properties	65.2	27
Washington, DC		
Mount Pleasant	67	18
Capitol Hill	72	15

Sources: Baltimore City Department of Housing and Community Development 1977; Gale 1976, 1977

In Philadelphia and elsewhere an "urban renaissance" of sorts may well have begun in the 1950s and 1960s, but it was not fueled by any significant return of the middle class from the suburbs. Even at the height of 1980s gentrification, suburban expansion proceeded apace. This would seem to cast doubt on the traditional cultural and economic explanations of gentrification as the result of altered consumer choices amid economic constraints. It is not that consumer choice is unimportant; in one scenario, it is possible that some gentrification involves younger people who moved to the city for an education and professional training in the decades after the 1950s but who did not then follow their parents' migration to the suburbs, becoming instead a social reservoir from which the gentrifier demand grew. If a dimension of consumer choice certainly remains, consumer *sovereignty* is more difficult to defend as a definitive explanation for gentrification. The problem is that gentrification is not simply a North American phenomenon but also emerged in the 1950s and 1960s in Europe and Australia (see, for example, Glass 1964; Pitt 1977; Kendig 1979; Williams 1984b, 1986), where the extent and experience of prior middle-class (and indeed working-class) suburbanization and the relation between suburb and inner city are substantially different. Only Ley's (1978) more general societal hypothesis about postindustrial cities is broad enough to account for the process internationally while retaining a consumption-centered approach, but the implications of accepting this view are somewhat drastic. If cultural choice and consumer preference really explain gentrification, this amounts either to the hypothesis that individual preferences change in unison not only nationally but internationally – a bleak view of human nature and cultural individuality – or that the overriding constraints are strong enough to obliterate the individuality implied in consumer preference. If the latter is the case, the concept of consumer preference is at best contradictory: a process first conceived in terms of individual consumption preference has now to be explained as resulting from cultural unidimensionality in the middle class – still rather bleak. At best, then, a focus on consumption can be rescued as theoretically viable only if it is used to refer to collective social preference, not individual preference.

The broader critique of the theory and assumptions underlying traditional urban economic theory is now well known (Ball 1979; Harvey 1973; Roweis and Scott 1981). I want here just to consider one particular aspect of neoclassical theory as it is applied to neighborhood change, leading to gentrification. To explain contemporary changes in the inner-city housing market, Brian Berry among others resorts to a "filtering" model. According to this model, new housing is generally occupied by better-off families who vacate their previous, less spacious housing, leaving it to be taken by poorer occupants, and move out toward the suburban periphery. In this way, decent housing "filters" down and is left behind for lower-income families; the worst housing drops out of the market to abandonment or demolition (Berry 1980: 16; Lowry 1960). Leaving aside entirely the question whether this "filtering" in fact guarantees "decent" housing for the working class, the filtering model is clearly based on a historicization of the effects of consumer sovereignty. People possess a set of consumer preferences, including a preference for more and more residential space, the model assumes, and so the greater one's ability to pay for space, the more space one will purchase. Smaller, less desirable spaces are left behind for those less able to pay. Other factors certainly impinge on demand for housing as well as its supply, but this preference for space together with the necessary income constraints provide the foundation for neoclassical treatments of urban development.

Gentrification contradicts this foundation of assumptions. It involves a so-called filtering in the opposite direction and seems to contradict the notion that preference for space *per se* is what guides the process of residential development. This means either that this assumption should be dropped from the theory or that so-called "external factors" and income constraints were so altered as to render the preference for more space impractical and inoperable. It is in this way that gentrification is rendered an exception – a chance, extraordinary event, the accidental outcome of a unique mix of exogenous factors. But in reality gentrification is not so extraordinary; it is extraordinary only to the theory which assumes it impossible from the start. The experience of gentrification illustrates well the limitations of neoclassical urban theory since in order to explain the process, the theory must be abandoned, and a superficial explanation based on *ad hoc* external factors must be adopted. But a list of factors does not make an explanation. The theory claims to explain suburbanization but cannot at all explain the historical continuity from suburbanization to gentrification and inner-city gentrification. Berry implicitly recognizes the need for (but lack of) such historical continuity when he concludes:

> a restructuring of incentives played a critical role in the increase in home ownership and the attendant transformation of urban form after the Second World War. There is no reason to believe that another restructuring could not be designed to lead in other directions, for in a highly mobile market system nothing is as effective in producing change as a shift in relative prices. There is, then, a way. Whether there is a will is another matter, for under conditions of democratic pluralism, interest group politics prevail, and the normal state of such politics is "business as

usual." The bold changes that followed the Great Depression and the Second World War were responses to major crises, for it is only in a crisis atmosphere that enlightened leadership can prevail over the normal business of politics in which there is an unerring aim for the lowest common denominator. Nothing less than an equivalent crisis will, I suggest, enable the necessary substantial inner city revitalization to take place.

(Berry 1980: 27–28).

In this way Berry shares with optimistic proponents of the process a voluntarist explanation of gentrification.

This critique of the neoclassical assumptions implicit in much gentrification research is partial and far from exhaustive. What it suggests, however, is the need for a broader conceptualization of the process, for the gentrifier as consumer is only one of many actors participating in the process. To explain gentrification according to the gentrifier's preferences alone, while ignoring the role of builders, developers, landlords, mortgage lenders, government agencies, real estate agents – gentrifiers as producers – is excessively narrow. A broader theory of gentrification must take the role of the producers as well as the consumers into account, and when this is done it appears that the needs of production – in particular the need to earn profit – are a more decisive initiative behind gentrification than consumer preference. This is not to say in some naive way that consumption is the automatic consequence of production, or that consumer preference is a totally passive effect of production. Such would be a producer's sovereignty theory, almost as one-sided as its neoclassical counterpart. Rather, the relationship between production and consumption is symbiotic, but it is a symbiosis in which the movement of capital in search of profit predominates. Consumer preference and demand for gentrified housing can be and is created, most obviously through advertising. Even in such early projects as Society Hill, a Madison Avenue firm was hired to sell the project (Old Philadelphia Development Corporation 1970). Although they are of secondary importance in initiating the actual process, and therefore in explaining why gentrification occurred in the first place, consumer preference and demand are of primary importance in determining the final form and character of revitalized areas – the difference between Society Hill, say, and London's Docklands or Brisbane's Spring Hill.

The so-called "urban renaissance" has been stimulated more by economic than cultural forces. In the decision to rehabilitate an inner-city structure, one consumer preference tends to stand out through the others: the need to make a sound financial investment in purchasing a home. Whether or not gentrifiers articulate this preference, it is fundamental, for few would even consider rehabilitation if a financial loss were expected. A theory of gentrification must therefore explain why some neighborhoods are profitable to redevelop while others are not. What are the conditions of profitability? Consumer sovereignty explanations have taken for granted the availability of areas ripe for gentrification when this was precisely what had to be explained.

Alternative explanations will involve a more detailed understanding of the broader historical and structural context of capital investment in the built environment and its role in urban development.

INVESTMENT IN THE BUILT ENVIRONMENT

In a capitalist economy, land and the improvements built onto it become commodities. As such they boast certain idiosyncrasies of which three are particularly important for this discussion.

First, private property rights confer on the owner near-monopoly control over land and improvements, monopoly control over the uses to which a certain space is put. Certainly zoning, eminent domain and other state regulations put significant limits on the landowner's control of land, but in the capitalist economies of North America, Europe and Australia these limitations are rarely if ever severe enough to displace the market as the basic institution governing the transfer and use of land. From this condition derives the importance of ground rent as a means to organize the geography of economic location.

Second, land and improvements are fixed in space but their value is anything but fixed. Improvements on the land are subject to all the normal influences on their value but with one vital difference. On the one hand, the value of the built improvements on a piece of land, as well as on surrounding land, influences the ground rent that landlords can demand; on the other hand, since land and buildings on it are inseparable, the price at which buildings change hands also reflects the ground rent level. Meanwhile a piece of land, unlike the improvements built on it, "does not require upkeep in order to continue its potential for use" (Harvey 1973: 158–159).

Third, while land is permanent, the improvements built on it are not but generally have a very long turnover period in physical as well as value terms. Physical decay is unlikely to claim the life of most buildings for at least twenty-five years, usually a lot longer, and it may take as long in economic (as opposed to accounting) terms for the building to pay back its value. From this we can derive several things: in a well-developed capitalist economy, large initial outlays will be necessary for investments in the built environment; financial institutions will therefore play an important role in the urban land market (Harvey 1973: 159); and patterns of capital depreciation will be an important variable in determining whether and to what extent a building's sale price reflects the ground rent level. These points will be of central importance in understanding patterns of investment and disinvestment.

In the economy, profit is the gauge of success, and competition is the mechanism by which success or failure is translated into growth or collapse. All individual enterprises must strive for higher and higher profits to facilitate the accumulation of greater and greater quantities of capital in profitable pursuits. Otherwise they find themselves unable to afford more advanced production methods and therefore fall behind their competitors. Ultimately, this leads either to bankruptcy or a merger into a larger enterprise. This search for increased profits translates, at the scale of the whole economy, into the necessity of long-run economic growth; stability is synonymous with growth.

Particularly when economic growth is hindered elsewhere in the economy, or where profit rates are low, the built environment becomes a target for the switching of much profitable investment. This is particularly apparent with the experience of suburbanization; spatial expansion rather than expansion *in situ* was the response to the continual need for capital accumulation (Walker 1977; Harvey 1978). But suburbanization illustrates well the two-sided nature of investment in the built environment, for as well as being a vehicle for capital accumulation, it can also become a barrier to further accumulation. It becomes so by dint of the characteristics noted above: near-monopoly control of space; the fixity of investments; the long turnover period. Near-monopoly control of space by landowners may prevent the sale of land for new development; the fixity of investments forces new development to take place at other, perhaps less advantageous, locations, and prevents redevelopment from occurring until invested capital has lived out its economic life; the long turnover period of capital invested in the built environment can discourage investment as long as other sectors of the economy with shorter turnover periods remain profitable. The early industrial city presented just such a barrier by the later part of the nineteenth century, eventually prompting suburban development rather than development *in situ*.

During the nineteenth century, in most cities land values displayed something approximating the classical conical form: a peak at the urban center with a declining gradient on all sides toward the periphery. It is probably fair to say that while this conical rent gradient was certainly evident in Europe (Whitehand 1987: 30–70), it is best exemplified in North America and perhaps Australia, where industrialization took place if not *de novo* then at least in the context of a recently developed urban structure. There too the market was freer from state regulation. This, of course, was the pattern that Homer Hoyt (1933) found in Chicago. With continued urban development the land value gradient is displaced outward and upward; land at the center grows in value while the base of the cone broadens. Land values tend to change in unison with long cycles in the economy; they increase most rapidly during periods of particularly rapid capital accumulation and decline temporarily during slumps. And as Whitehand (1987: 50) has demonstrated for Glasgow, these different cycles of outward growth may bring different kinds and sources of built environment construction. Since suburbanization relied on considerable capital investments in land, construction, transportation, etc., it too tended to follow this cyclical trend. Faced with the need to expand the scale of their productive activities, and unable or unwilling for a variety of reasons to expand any further where they were, industries jumped out beyond the city to the base of the land value cone where extensive spatial expansion was most possible and relatively cheap. The alternative – substantial renewal and redevelopment of the already built up area – would have been too costly for private capital to undertake, and so industrial capital increasingly migrated to the new suburbs.

In the US this movement of industrial capital began in force after the severe depression of 1893–1897, somewhat later than in Europe's larger and older urban centers. It was both followed and paralleled by a substantial migration of capital for residential construction. In the already well-established cities, the only significant exception to this geographical refocusing of construction capital lay in the central business district (CBD), where substantial skyscraper office development began by the 1920s. In fact, the inner city was adversely affected by this movement of capital to the suburbs where higher returns were available: a combination of neglect and concerted disinvestment by investors, due to high risk and low rates of return, initiated a long period of deterioration and a lack of new capital investment in the inner city. In the words of a 1933 commentator:

> The simple fact is that while cities have continued to spread in an unprecedented manner, resulting in much financial embarrassment at the present time, their commercial and light-industrial areas [and working-class quarters] at the center have stopped spreading and in some cases show very definite signs of receding from former partially occupied boundaries.
>
> (Wright 1933: 417)

As a result, land values in the inner city generally fell relative to the CBD and the suburbs, and so by the late 1920s Hoyt could identify for Chicago a newly formed "valley in the land-value curve between the Loop and outer residential areas" (Figure 3.1). This valley "indicates the location of those sections where the buildings are mostly over forty years old and where the residents rank lowest in rent-paying ability" (Hoyt 1933: 356–358). Hoyt noted this oddity – an apparent aberration in the conical rent gradient – and puzzled about it, but moved on. In fact, throughout the decades of most sustained suburbanization, from the 1940s to the 1960s, this valley in the land value curve deepened and broadened owing to a continued lack of productive capital investment.

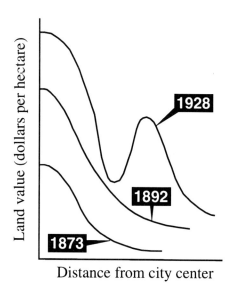

Figure 3.1 The ground rent surface and evolution of the land value valley in Chicago (after Hoyt 1933)

By the late 1960s the valley may have been as much as six miles wide in Chicago (McDonald and Bowman 1979), and a similar size in New York City (Heilbrun 1974: 110–111). Evidence from other cities suggests that this capital devalorization and consequent broadening of the land value valley occurred in most older cities in the US (Davis 1965; Edel and Sclar 1975), producing the slums and ghettos that were suddenly discovered as "problems" in the postwar era by the departed suburban middle class.

A theory of gentrification will need to explain the historical process of capital devalorization in the inner city and the precise way in which this devalorization produces the possibility of profitable *re*investment. The crucial nexus here is the relationship between land value and property value. As they stand, however, these concepts are insufficiently refined. Land value for Hoyt was a composite category referring to the price of undeveloped plots and the expected future income from their use; the type of future use was simply assumed. Property value, on the other hand, is generally taken to mean the price at which a building is sold, and thus includes the value of the land. To elaborate the relationship between land value and the value of buildings in fuller detail, then, it will be necessary to disaggregate these two measures of value into four separate but related categories. These four categories (house value, sale price, capitalized ground rent, potential ground rent) remain fully or partially obscure and indistinguishable under the umbrella concepts land value and property value.

House value

Consistent with its emphasis on consumer preference, neoclassical economic theory explains prices as the result of supply and demand conditions. But if, as suggested above, the search for a high return on productive investments is the primary initiative behind gentrification, then the specific costs of production (not just the quantity of end-product – i.e. supply) will be central in the determination of prices. In opposition to neoclassical theory, therefore, it will be necessary to separate the value of a house from its price. Following the classical political economists (Smith, Ricardo), and after them Marx, I take as axiomatic a labor theory of value: the value of a commodity is measured by the quantity of socially necessary labor power required to produce it. Only in the marketplace is value translated into price. And although the price of a house reflects its value, the two cannot mechanically be equated since price (unlike value) is also directly affected by supply and demand conditions. Thus value considerations (the amount of socially necessary labor power performed in making the commodity) set the level about which the price fluctuates. Now with housing, the situation is further complicated insofar as individual houses return periodically to the market for resale. The house's value will also depend, therefore, on its rate of devalorization through use, versus its rate of revalorization through the addition of more value. The latter occurs when further labor is performed for maintenance, replacement, extensions, etc.

Sale price

A further complication with housing is that the sale price represents not only the value of the house, but an additional component for rent since the land is generally sold along with the structures it accommodates. Here it is preferable to talk of ground rent rather than land value, since the price of land does not reflect a quantity of labor power applied to it, as with the value of commodities proper.

Capitalized ground rent

Ground rent is a claim made by landowners on users of their land; it represents a reduction from the surplus value created over and above cost price by producers on the site. Capitalized ground rent is the actual quantity of ground rent that is appropriated by the landowner, given the present land use. In the case of rental housing where the landlord produces a service on land he or she owns, the production and ownership functions are combined and ground rent becomes even more of an intangible category though nevertheless a real presence; the landlord's capitalized ground rent returns mainly in the form of house rent paid by the tenants. In the case of owner-occupancy, ground rent is capitalized only when the building is sold and therefore appears as part of the sale price. Thus, assuming for the moment an equation between price and value, sale price = house value + capitalized ground rent.

Potential ground rent

Under its present land use, a site or neighborhood is able to capitalize a certain quantity of ground rent. For reasons of location, usually, such an area may be able to capitalize higher quantities of ground rent under a different land use. Potential ground rent is the amount that could be capitalized under the land's "highest and best use" (in planners' parlance) – or at least under a higher and better use. This concept is particularly important in explaining gentrification.

On the basis of these concepts, the historical process that has made certain neighborhoods ripe for gentrification can be outlined.

CAPITAL DEVALORIZATION IN THE INNER CITY

The physical deterioration and economic devalorization of inner-city neighborhoods is a strictly logical, "rational" outcome of the operation of the land and housing markets. This is not to suggest it is at all natural, however, for the market itself is a social product. Far from being inevitable, neighborhood decline is

> the result of identifiable private and public investment decisions. . . . While there is no Napoleon who sits in a position of control over the fate

of a neighborhood, there is enough control by, and integration of, the investment and development actors in the real estate industry that their decisions go beyond a response and actually shape the market.

(Bradford and Rubinowitz 1975: 79)

What follows is a rather schematic attempt to explain the historical decline of inner-city neighborhoods in terms of the institutions, actors and economic forces involved. We might think of this explanation as a production-side corrective to traditional "filtering" theory. It requires the identification of a few salient processes that characterize the different stages of decline, but it is not meant as a definitive description of what every neighborhood experiences. The day-to-day dynamics of decline are complex and, as regards the relationship between landlords and tenants in particular, have been examined in considerable detail elsewhere (Stegman 1972). This schema is, however, meant to provide a general explanatory framework within which each neighborhood's concrete experience can be understood. It is assumed from the start that the neighborhoods concerned are relatively homogeneous as regards the age and quality of housing, and, indeed, this tends to be the case with areas experiencing redevelopment.

New construction and the first cycle of use

When a neighborhood is newly built the price of housing reflects the value of the structure and improvements put in place plus the enhanced ground rent captured by the landowner. During the first cycle of use, the ground rent is likely to increase as urban development continues outward, and the house value will only very slowly begin to decline if at all. The sale price therefore rises. But eventually sustained devalorization of neighborhood housing can take hold, and this has three sources: advances in the productiveness of labor; style obsolescence; and physical wear and tear. Advances in the productiveness of labor are chiefly due to technological innovation and changes in the organization of the work process. These advances allow a similar structure to be produced at a lower value than was previously possible. Truss frame construction and the factory fabrication of parts in general, rather than on-site construction, are only two recent examples of such advances. Style obsolescence is secondary as a stimulus for sustained depreciation in the housing market and may even occasionally induce an appreciation in house prices insofar as older styles are more sought after than the new. Physical wear and tear also affects the value of housing, but it is necessary here to distinguish between minor repairs which must be performed regularly if a house is to retain its value (for example, painting doors and window frames, interior decorating) and major repairs which are performed less regularly but require greater outlays (for example, replacing the plumbing or electrical systems), and structural repairs without which the structure becomes unsound (for example, replacing a roof, or replacing floorboards that have dry rot). Devalorization of a property after one cycle of use reflects the imminent need not only for regular, minor repairs but also for a succession of more

Plate 3.1 Disinvestment in the urban housing stock: an abandoned building and community-based rehabilitation (*The Shadow*)

major repairs involving a substantial investment. Devalorization will induce a price decrease relative to new housing but the extent of this overall decrease will depend on how much the ground rent has also changed in the meantime.

Landlordism and homeownership

Clearly, property owners in many neighborhoods succeed in making major repairs and maintaining or even enhancing the value of an area's housing. These areas remain stable. Equally clearly, there are areas of owner-occupied housing which experience the first stage of devalorization. Homeowners, aware of imminent decline unless repairs are made, may sell out and seek newer homes where their investment will be safer. At this point, after a first or subsequent cycle of use, there is a tendency for the neighborhood to convert toward a higher level of rental tenancy unless repairs are made. And since landlords use buildings for different purposes than owner-occupiers, a different pattern of maintenance will ensue. Owner-occupiers in the housing market are simultaneously both consumers and investors; as investors, their primary return comes as the increment of sale price over purchase price. The landlord, on the other hand, receives his or her return mainly in the form of house rent, and under certain conditions may have a lesser incentive for carrying out repairs so long as he or she can still command rent. This is not to say that landlords typically undermaintain properties they possess; newer apartment complexes and even older accommodations for which demand is high may be very well maintained. But as Ira Lowry has indicated, "under-maintenance is an eminently reasonable response of a landlord to a declining market" (1960: 367), and since the transition from owner-occupancy to tenancy is generally associated with a declining market, some degree of undermaintenance could be expected.

Undermaintenance frees up capital that can be invested elsewhere. It may be invested in other city properties, it may follow developers' capital out to the suburbs, or it may be invested in some other sector of the economy entirely. With sustained undermaintenance in a neighborhood, however, it may become difficult for landlords to sell their properties, particularly since the larger financial institutions will now be less forthcoming with mortgage funds; sales become fewer and more expensive to the landlord. Thus, there is even less incentive to invest in the area beyond what is necessary to retain the present revenue flow. This pattern of decline is likely to be reversed only if a shortage of higher-quality accommodations occurs, allowing rents to be raised and making improved maintenance worthwhile. Otherwise, the area is likely to experience a net outflow of capital, which will be small at first since landlords still have substantial investments to protect. Under these conditions it becomes very difficult for the individual landlord or owner to struggle against the economic decline which they have helped to induce. House values are falling and the levels of capitalized ground rent for the area are dropping below the potential ground rent (see Figure 3.2). The individual who did not undermaintain his property would be forced to charge higher than average rent for the area with little hope of attracting tenants earning higher than average income which would capitalize the full ground rent. This is the celebrated "neighborhood effect" which operates through the ground rent structure.

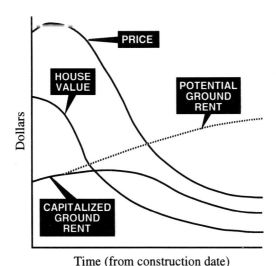

Figure 3.2 The devalorization cycle and the evolution of the rent gap

Blockbusting and blowout

Some neighborhoods may not transfer to rental tenancy and they will experience relative stability or a gentler continuation of decline. If the latter occurs, the owner-occupants may undermaintain, though usually out of financial constraints rather than market strategy. With blockbusting, this decline is intensified. Real estate agents exploit racist sentiments in white neighborhoods that are experiencing declining sale prices; they buy houses relatively cheaply, and then resell at a considerable markup to African-American, Latino or other "minority" families, many of whom may be struggling to own their first home. As Laurenti's research suggests, property values are usually declining before blockbusting takes place and do not begin declining simply as a result of racial changes in ownership (Laurenti 1960). Once blockbusting has taken place, however, further decline in house values is likely, not just because of the racism of the housing market but also because of the inflated prices at which houses were sold and the consequent lack of resources for maintenance and mortgage payments suffered by incoming families. Blowout, a similar process, operates without the helping hand of real estate agents. Describing the process as it operated in the Baltimore housing market during the 1960s, Harvey *et al.* (1972; see also Harvey 1973: 173) point to the outward spread of slums from the inner city (the broadening of the land value valley) and the consequent squeezing of still healthy outer neighborhoods against secure upper-middle-class residential enclaves lying further out. Thus squeezed, owner-occupants in an entire neighborhood are likely to sell out, often to landlords, and move to the suburbs.

Redlining

Undermaintenance gives way to more active disinvestment as capital depreciates further and the landlord's stake diminishes; house value and capitalized ground rent fall, producing further decreases in sale price. Disinvestment by landlords is accompanied by an equally "rational" disinvestment by financial institutions, which cease supplying mortgage money to the area. Larger institutions offering low-downpayment, low-interest-rate loans find they can make higher returns in the suburbs with a lower chance of foreclosure and less risk of declining property values. Their role in the inner city is taken over initially by smaller, often local organizations specializing in higher-risk financing. Redlined by larger institutions, the area may also receive loans insured by the Federal Housing Administration (FHA), although these too were virtually confined to the outer city. Though meant to prevent decline, FHA loans have even contributed to decline in places (Bradford and Rubinowitz 1975: 82). In addition to mortgage redlining, there is also redlining on the part of homeowner insurance companies (Squires *et al.* 1991), which further induces economic disinvestment. What loans do occur at this stage allow properties to change hands but do little to encourage reinvestment in maintenance so the process of decline can simply be lubricated. Ultimately, medium and small-scale investors also refuse to work the area, as do mortgage insurers.

Vandalism further accelerates devalorization and becomes a problem especially when properties are temporarily vacant between tenants (Stegman 1972: 60). Even when a building is occupied, however, vandalism may contribute to devalorization, especially if it is being undermaintained or systematically "milked". Vandalism is actually a landlord strategy at this stage, whether in New York (Salins 1981) or, less commonly, in London (Counsell 1992). Subdivision of structures to yield more rental units is common at this stage. By subdividing, the landlord hopes to intensify the building's use (and profitability) in its last few years. But eventually landlords will disinvest totally, refusing to make repairs and paying only the necessary costs – and then often only sporadically – for the building to yield rent.

Abandonment

When landlords can no longer collect enough house rent to cover the necessary costs (utilities and taxes), buildings are abandoned. This is a neighborhood-scale phenomenon; the abandonment of isolated properties in otherwise stable areas is rare. Much abandoned housing is structurally sound, which seems paradoxical. But then buildings are abandoned not because they are unusable, but because they cannot be used *profitably*. At this stage of decline, there is a certain incentive for landlords to destroy their own property through arson and collect the substantial insurance payment.

GENTRIFICATION – THE RENT GAP

The previous section presented a summary sketch of the process which was commonly but misleadingly referred to in the 1960s and 1970s as "filtering." It is a common process in the housing market and affects many neighbor-hoods; if it is most accomplished and most clearly evident in the cities of the US, it is not exclusively a US phenomenon, as Friedrichs' (1993) comparative research on Germany and the US makes clear. But by the same token, this cycle of devalorization is by no means universal, nor does it take place in pre-cisely the same manner in every neighborhood. It is included here precisely because gentrification is generally preceded by such a cycle, although the process need not occur fully for gentrification to ensue. Nor should this decline be thought of as inevitable. As Lowry quite correctly insists, "filtering" is due not simply "to the relentless passage of time" but to "human agency" (1960: 370). The previous section has suggested who some of these agents are, and the market forces they both react to and help create. It also suggests that the objective mechanism underlying filtering is the depreciation and devalorization of capital invested in residential inner-city neighborhoods. This devalorization produces the objective economic conditions that make capital *revaluation* (gentrification) a rational market response. Of fundamental importance here is what I call the rent gap.

The rent gap is the disparity between the potential ground rent level and the actual ground rent capitalized under the present land use (Figure 3.2). The rent gap is produced primarily by capital devalorization (which diminishes the proportion of the ground rent able to be capitalized) and also by continued urban development and expansion (which has historically raised the potential ground rent level in the inner city). The valley which Hoyt detected in his 1928 observation of land values (Figure 3.1) can now be understood in large part as the result of a developing rent gap. Only when this gap emerges can reinvestment be expected since if the present use succeeded in capitalizing all or most of the ground rent, little economic benefit could be derived from redevelopment. As filtering and neighborhood decline proceed, the rent gap widens. Gentrification occurs when the gap is sufficiently wide that developers can purchase structures cheaply, can pay the builder's costs and profit for rehabilitation, can pay interest on mortgage and construction loans, and can then sell the end product for a sale price that leaves a satisfactory return to the developer. The entire ground rent, or a large portion of it, is now capitalized; the neighborhood is thereby "recycled" and begins a new cycle of use.

We have focused here on the widespread situation in which a cycle of capital devalorization, brought about by disinvestment, accounts for the emergence of a rent gap. But it is also possible to conceive of a situation in which, rather than the capitalized ground rent being pushed down through devalorization, the potential ground rent is suddenly pushed higher, opening up a rent gap in a different manner. This might be the case, for example, when there is rapid and sustained inflation, or where strict regulation of a land market keeps potential ground rent low, but is then repealed. This contribution to the formation of a rent gap might be significant in explaining gentrification in Amsterdam and Budapest (see Chapter 8).

Once the rent gap is wide enough, gentrification may be initiated in a given neighborhood by any of several different actors in the land and housing market. And here we come back to the relationship between production and consumption, for the empirical evidence suggests that as often as not, the process is initiated not by the exercise of those individual consumer preferences much beloved of neoclassical economists, but by some form of *collective social action* at the neighborhood level. The state, for example, initiated much of the early gentrification in the US as a continuation of urban renewal projects, and though it plays a lesser role today, state subsidies and sponsorship of gentrification remain important. More commonly today, with private-market gentrification, one or more financial institutions will reverse a long-standing redlining policy and actively target a neighborhood as a potential market for construction loans and mortgages. All the consumer preference in the world will amount to naught unless this long-absent source of funding reappears; mortgage capital, in some form or other, is a prerequisite. Of course, this mortgage capital must be borrowed by willing consumers exercising some preference or another. But these preferences can be and are to a significant degree socially created. Along with financial institutions, professional developers have generally acted as the collective initiative behind gentrifica- tion. Typically, a developer will purchase not one but a significant proportion of devalorized properties in a neighborhood for rehabilitation and sale. A significant exception to this predominance of collective capital in the initiation of gentrification occurs in neighborhoods adjacent to already gentrified areas.

There, indeed, it is common to find that individual gentrifiers may be very important in initiating rehabilitation. Their decision to rehabilitate followed the results from the previous neighborhood, however, which implies that sound financial investment was uppermost in their minds. And they still require mortgage capital from willing institutions.

Three kinds of developers typically operate in recycling neighborhoods: (a) professional developers who purchase property, redevelop it, and resell for profit; (b) occupier developers who buy and redevelop property and inhabit it after completion; and (c) landlord developers who rent to tenants after rehabilitation.[2] The developer's return on investment comes as part of the completed property's sale price; for the landlord developer it also comes in the form of house rent. Two separate gains comprise the return achieved through sale: capitalization of enhanced ground rent, and profit (quite distinct from builder's profit) on the investment of productive capital. Professional and landlord developers are important – contrary to the public image, they were by far the majority in Society Hill – but occupier developers are more active in rehabilitation than they are in any other sector of housing construction. Since the land has already been developed and an intricate pattern of property rights laid down, it is not always easy for the professional developer to assemble sufficient land and properties to make involvement worthwhile. Even landlord developers tend to be rehabilitating several properties simultaneously or in sequence. The fragmented structure of property ownership has made the occupier developer, who is generally an inefficient operator in the construction industry, into a plausible vehicle for remaking devalorized neighborhoods.

Viewed in this way, gentrification is not a chance occurrence or an inexplicable reversal of some inevitable filtering process. On the contrary, it is to be expected. The devalorization of capital in nineteenth-century inner-city neighborhoods, together with continued urban growth during the first half of the twentieth century, have combined to produce conditions in which profitable reinvestment is possible. If this rent gap theory of gentrification is correct, it might be expected that rehabilitation began where the gap was greatest and the highest returns available, that is, in neighborhoods particularly close to the city center and in neighborhoods where the sequence of declining values had pretty much run its course. But too much could be made of this expectation. Empirically, gentrification has indeed tended to hug the city center, at least in the early stages, but too much goes into the immediate causes of gentrification in a particular neighborhood for it to be possible to correlate level of decline with propensity to gentrify. The theory would also suggest that as these first areas are recycled, other areas offering lower but still substantial returns – or areas presenting fewer obstacles to reinvestment – would be sought out by developers. This would involve areas farther from the city center and areas where decline was less advanced. Thus in Philadelphia, South Street, Fairmount and Queen Village became the new "hot spots" following Society Hill (Cybriwsky 1978; Levy 1978), and the city's triage policy for allocating block grant funds made other parts of near North and West Philadelphia likely candidates for future reinvestment in some form or another.

The state's role in earlier rehabilitation schemes is worth noting. By assembling properties at a "fair market value" and returning them to developers at the lower assessed price, the state accomplished and bore the costs of the last stages of capital devalorization, thereby ensuring that developers could reap the high returns without which rehabilitation or redevelopment would not occur. With the state now less involved in this writing down of property values, developers are clearly able to absorb the costs of devaluing capital that has not yet fully devalorized. That is, they can pay a relatively high price for properties to be rehabilitated, and still make a reasonable return. It seems, then, that whatever social and political failures accompanied urban renewal – and they were of course many – the state has actually been successful in economic terms insofar as it has provided the broad conditions that would stimulate private market revitalization.

CONCLUSION:
A BACK-TO-THE-CITY MOVEMENT BY CAPITAL

Gentrification is a structural product of the land and housing markets. Capital flows where the rate of return is highest, and the movement of capital to the suburbs, along with the continual devalorization of inner-city capital, eventually produces the rent gap. When this gap grows sufficiently large, rehabilitation (or, for that matter, redevelopment) can begin to challenge the rates of return available elsewhere, and capital flows back in. Gentrification is a back-to-the-city movement all right, but a back-to-the-city movement by capital rather than people.

The advent of gentrification in the latter part of the twentieth century has demonstrated that contrary to the conventional neoclassical wisdom, middle- and upper middle-class housing can be intensively developed in the inner city. Gentrification itself has now significantly altered the urban ground rent gradient. The land value valley may be being displaced outward and in part upward as gentrification revalues central city land (Figure 3.3), and as disinvestment is displaced outward to the closer, older suburbs leading in turn to a new flurry of complaints that middle-class suburbs now face "city problems" (Caris 1996; Schemo 1994).

DOONESBURY   **by Garry Trudeau**

Plate 3.2 Dr. Dan explains gentrification (Doonesbury © 1980 G. B. Trudeau. Reprinted with permission of Universal Press Syndicate. All rights reserved)

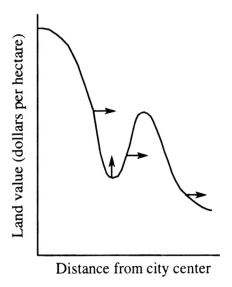

Figure 3.3 Evolution of the ground rent surface and land value valley following gentrification

Gentrification has been the leading residential and recreational edge (but in no way the cause) of a larger restructuring of space. At one level, restructuring is accomplished according to the needs of capital, accompanied by a restructuring of middle-class culture. But in a second scenario, the needs of capital might be systematically dismantled, and a more social, economic and cultural agenda addressing the direct needs of people might be substituted as a guiding vision of urban restructuring.

In the meantime, however, it is difficult to speculate much further about the immediate prospects for gentrification and the city, solely from the local perspective adopted in this chapter. The process, after all, is tightly bound up with the patterns and rhythms of capital investment in the built environment and of capital accumulation and crisis more generally.

POSTSCRIPT

Since the rent gap theory was originally proposed in 1979, it has become the subject of considerable debate and research in the urban literature. Predictably, perhaps, it has been attacked for displacing consumer preference and individual choice from its explanatory pedestal and replacing individual consumption with the movement of capital (Ley 1986; Mills 1988; Caulfield 1989, 1994; Hamnett 1991; response by Clark 1992). Some have suggested an ambivalent connection between rent gap theory and Marshallian economics (Clark 1987) while others have tried to gentrify the theory itself back into safe if confused neoclassical terms (Bourassa 1993, and response by Clark 1995; Boyle 1995; N. Smith 1995b). More directly, some have simply denied the existence of anything approximating a rent gap (Ley 1986, and for a response, N. Smith 1987; Bourassa 1990, and responses by Badcock 1990). More reasonably, a number of critics have pointed to the limits of the theory. Beauregard (1990) argues that the rent gap theory cannot predict precisely which neighborhoods will gentrify and which will not, and Badcock (1989) points out that gentrification (in the narrower sense of residential rehabilitation) was actually a third choice for filling in the rent gap in Adelaide. The theory also omits the clear connections to social change that come with gentrification, and in particular it does not explain the emergence of the agents of gentrification (D. Rose 1984, 1987; Beauregard 1986, 1990).

The latter criticisms have some merit and I think they serve to establish some limits to the applicability of the rent gap theory, which was, after all, deliberately intended to view gentrification through the lenses of the local housing market. It is also a difficult concept to render operational (Ley 1986; N. Smith 1987). Nonetheless, since the initial proposal of the theory, rent gaps have been identified in a number of cities undergoing gentrification. Identification of the rent gap depends on finding appropriate measures for capitalized and potential ground rent, and in different national contexts different sources have been used for this purpose.

Clark's (1987) landmark study painstakingly identified the rent gap in several sampled blocks in central Malmö in Sweden. A significant rent gap began to emerge toward the end of the nineteenth century in Malmö but began to close with gentrification activity and redevelopment in the late 1960s and early 1970s (see also Clark 1988). More recently Clark and Gullberg (1991) have examined the interaction between rent gaps, long swings in urban building and different forms of building in Stockholm. An equally meticulous study by Engels (1989) was able to measure the evolution of a rent gap in Glebe, a suburb of Sydney, beginning in the earlier years of this century. Here too there is evidence of a considerable closing of the gap in the early 1970s with the onset of gentrification. Both studies suggest that speculation ahead of physical reconstruction can also significantly diminish the rent gap.

While questioning other aspects of gentrification, Badcock (1989) was able to identify the rent gap in Adelaide while pointing out that a number of different strategies of revaluing a devalorized landscape might well be adopted

DOONESBURY   **by Garry Trudeau**

Plate 3.3 Dr. Dan explains the rent gap (Doonesbury © 1980 G. B. Trudeau. Reprinted with permission of Universal Press Syndicate. All rights reserved)

depending on local conditions and government initiatives (see also Badcock 1992a, 1992b). Allison identified the "valley in land values" in Brisbane's Spring Hill but found it difficult to quantify precisely (Allison 1995: 165). Kary (1988) has charted a land value valley around Toronto in the 1960s and early 1970s but notes that the depth of the valley is far from uniform around the city. He goes on to identify the rent gap in his case study of the Cabbagetown/Donvale district and he too traces the evidence of infilling following gentrification in the late 1970s and 1980s. Cortie and van de Ven (1981) and van Weesep and Wiegersma (1991) mark the presence of rent gaps in Amsterdam.

In the context of the London property market, Hamnett and Randolph (1984, 1986) identify what they call a "value gap" between the "vacant possession value" of a property and its "tenanted investment value." Where the "value gap" becomes sufficiently large, the property owner is encouraged to transfer the building from rental residential to other tenure forms. A couple of points about this argument are relevant here. In the first place, if we are to be consistent with the distinction between price and value, the value gap would more properly be referred to as a "price gap." Given its broad currency, however, it is probably not useful to insist on this pedantic nomenclature. Second, there is clearly a relationship between the "value" gap and the rent gap. As Clark (1991a) points out, the rent gap theory does not directly address the question of tenure conversion in the process of gentrification, and so the value gap can be seen as a complementary refinement of the rent gap argument. As Clark concludes, "a property will not have a value gap without also having a rent gap" (Clark 1991b: 24).

Finally a historical note. Clark (1987) is surely correct to argue that although the rent gap theory sounds very novel its antecedents lie in both Friedrich Engels *and* Alfred Marshall. And others surely also anticipated the idea. A 1933 account envisions the rent gap explanation of gentrification and redevelopment to an astonishing degree.[3] It may be a "startling claim," begins regional planner Henry Wright, but "the greatest impediment to slum clearance, the high cost of land near the city center is already well on the way to dissolution." He goes on to identify "actual and potential land values" and draws a chart of their divergence:

> Thus we find actual use values for the slum area . . . subject to reduction by a double pull, outward to the new suburbs . . . and the new and scarcely recognized pull inward toward the skyscraper center of reducing capacity requirements. The slum is left an "orphaned" district. . . . But these usually astute [real estate] interests have as yet failed to acknowledge the full losses in a shrinkage to a final "real value" for the only purpose for which there remains a probable use: their reconstruction for residential purposes and this to be of a kind capable of absorbing relatively large areas of land.
>
> (Wright 1933: 417–418)

In other words, they have not yet perceived the "potential land value," in Wright's words, that could come from renovation and redevelopment.

Having identified the problem, Wright is equally direct about the solution. These "nearer areas," he says "should be properly made over on the basis of land costs":

> I do not hesitate to say at once that the idea that the slums should be rebuilt primarily with the vision of rehousing the present tenants is no longer valid in respect to any large-scale handling of the problem. . . . why should we not take advantage of the situation to readjust our ideas about desirable dwelling ideas and recreate the present slum districts for the convenient and enjoyable occupancy of those whose business relations are largely in the central area?
>
> (Wright 1933: 417, 419)

If it had to endure the failed agenda of urban renewal – indeed ironically, was enhanced by it – this agenda for a "make over on the basis of land costs" (Wright 1933: 417), dating back to the 1930s, surely found perfect expression in the gentrification of recent decades.

4

GLOBAL ARGUMENTS

Uneven development

Gentrification is the product of local housing markets, and for that reason I tried in the preceding chapter to begin to theorize the process largely at the local scale. The rent gap theory speaks to the relationship between individual structures and lots and neighborhood-scale dynamics in the land and housing markets; it involves a knowledge of specific actors in these markets, and refers to the history of investment and disinvestment at the neighborhood scale. But in addition to these local dynamics, gentrification represents a significant historical geographical reversal of assumed patterns of urban growth intimately connected to a wider frame of political-economic change. Gentrification occurs in cities in at least three continents, and is closely connected with what came in the 1980s to be seen as "globalization." We need, then, to come at gentrification from the other side as well, from its position in the global economy. This is perhaps best achieved by trying to understand gentrification in terms of the "uneven development" of the global and national economies.

The wider purview on gentrification is important because it helps to address a central question in the public as well as academic debate about urban change. How significant is gentrification as a fabricant of the urban landscape? For some, gentrification is a localized, small-scale process which, while maybe symbolically important, is purely temporary and of little long-term significance. Gentrification is a short-lived exception. This is the position that the prominent urban geographer Brian Berry, for example, has consistently taken: gentrification, he urges, should be seen as small "islands of renewal in seas of decay" (Berry 1973, 1980, 1985). If the surge of gentrification in the 1980s kept this exceptionalist position somewhat at bay, it is also the position that was quickly adopted by real estate professionals in the late 1980s in the US and Europe at the end of the economic boom of the 1980s. The reasoning here is that the particular factors combining to encourage gentrification are themselves purely temporary: the high cost of suburban housing, low housing vacancy rates, lifestyle changes in the baby boom generation, yuppie consumption habits do not represent long-term shifts, and when they cease to operate, gentrification too will cease.

By contrast, other urban commentators have seen in gentrification a long-term urban reversal. Gentrification in this vision may represent only part of a larger "revitalization" of the city, a recentralization of specific urban activities over and against the suburbs. It is lauded as part of a spontaneous recentralization of services, recreational facilities and employment opportunities as well as elite housing. In its most optimistic renderings, gentrification is seen as part of a larger economic shift and social movement that has the potential to reverse the historic decline of the central and inner city (see for example Laska and Spain 1980). Yet whatever its rosy optimism, this hope of a gentrified future is rarely based explicitly on any broader explanatory perspective. Its grounding in optimism more than theory reached a peak when one of its better-known adherents, Jimmy Carter, chose the South Bronx to symbolize decay that could be reversed; more than knowledge and understanding, it was hope and belief in the long-term salutary effects of gentrification that motivated President Carter's sojourn to the South Bronx and much of the stillborn National Urban Policy that followed.

If theories of gentrification are more assumed than explicated in the latter optimistic vision, one thing is striking. The assumed causes of this reversal are really quite similar to the exceptionalist position, despite the fact that they seem to explain opposite conclusions. Both treat urban change as driven by consumption-side considerations; the dispute is merely about the extent to which consumption choices might change. But how can a consistent urban theory lead to directly opposite conclusions? Berry is surely correct to argue that in their enthusiasm for an apparently novel development, the optimists have ignored or undervalued the counsel of hard, tested theories. But, as we saw in the last chapter, Berry's own position is not unproblematic. Adhering most consistently to the traditional urban economic theories, Berry's voluntarist explanation leads him to dismiss the extent and significance of gentrification. Recall that for Berry, gentrification would have to be the result of "a restructuring of incentives," a process he concedes is possible but would take a bold civic vision. In fact, Berry argues (pp. 56–57, above), such a change in incentives and therefore consumption patterns would take place only if the structure of constraints is changed: "The bold changes that followed the Great Depression and the Second World War" achieved such a shift, he suggests, and something at the same scale would have to be replicated now if gentrification were to become significant. These bold changes represented "responses to major crises, for it is only in a crisis atmosphere that enlightened leadership can prevail over the normal business of politics in which there is an unerring aim for the lowest common denominator. Nothing less than an equivalent crisis," Berry suggested in 1980, would "enable the necessary substantial inner city revitalization to take place" (Berry 1980: 27–28).

In retrospect, by 1980 we were already well into just such a crisis – not just nationally but internationally, not just in the residential sector but throughout the economy – and this crisis did indeed begin to realize a lot more than simply a restructuring of "prices" and "incentives" (Harris 1980a; Massey and Meegan 1978). But crises and restructuring are not exogenous "factors," accidental departures from equilibrium, as is generally assumed in neoclassical theory. Economic crises are concrete historical events which, as well as throwing up new situations and social relationships, realize in a short period a number of tendencies already implicit in the economy (Harman 1981; Harvey 1982). In short, a restructuring of urban space has been afoot since the 1970s, and while this restructuring certainly involves such "factors" as the baby boom, energy prices and the cost of new housing units, its roots and its momentum derive from a deeper and very specific set of processes that we can refer to as uneven development. At the urban scale, gentrification represents the leading edge of this process.

UNEVEN DEVELOPMENT AT THE URBAN SCALE

By uneven development is often meant the self-evident truth that societal development does not take place everywhere at the same speed or in the same direction. Such an obvious notion would barely deserve mention, far less the scrutiny it has received. Rather, uneven development should be conceived as a quite specific process that is both unique to capitalist societies and rooted directly in the fundamental social relations of this mode of production. To be sure, societal development in other modes of production may well be uneven, but it is so for quite different reasons, has a different social significance, and results in different geographical landscapes. The geography of the feudal market town is systematically different from that of the contemporary metropolis. Under capitalism the relationship between developed and under-developed areas is the most obvious and most central manifestation of uneven development, and occurs not just at the international scale but also at regional and urban scales (Soja 1980). At different spatial scales, capital moves geographically for different but parallel reasons, and it is this parallelism of purpose and structure that engenders a similar spatial unevenness at different scales. Here it is possible only to sketch part of the economic rationale of uneven development, and to do so in the most summary fashion (N. Smith 1984). I will take three central aspects of uneven development and, by examining them sequentially, I hope to piece together a framework for the theory. At each step, I will locate gentrification within the analysis, thereby providing an illustration for uneven development theory as well as a broader theoretical framework within which to understand gentrification.

Tendencies toward differentiation and equalization

Inherent in the structure of capitalism are two contradictory tendencies toward, on the one hand, the *equalization of conditions and levels of development* and, on the other, *their differentiation*. The tendency toward equalization emerges from the more basic necessity for economic expansion in capitalist society: individual capitalists and enterprises can survive only by making a profit, but in an economy ruled by competition between separate enterprises, survival requires expansion – the accumulation of larger and larger quantities of capital. At the level of the national or world economy, this translates into the necessity of permanent economic growth; when such growth does not occur, the system is in crisis. Economic expansion is fueled by drawing more and more workers into waged labor and productive consumption, by locating

Plate 4.1 From the "armpit of America" to "Charm City":
the rebuilt Baltimore waterfront

and exploiting increased quantities of raw materials, and by developing
the means of transportation that provide cheaper and faster access to raw
materials and markets. In short, expansion is fueled by creating a larger
number and broader variety of commodities, by selling them on the market,
and by reinvesting part of the profit in a further expansion of the scale of
the productive forces. Historically, the earth is transformed into a universal
means of production, and no corner is immune from the search for raw
material; the land, the sea, the air and the geological substratum are reduced
in the eyes of capital to a real or potential means of production, each with
a price tag. This is the process that lies behind the tendency toward an
equalization of levels and conditions of production. Thus it is that a new car
plant in Tokyo is much the same as a new car plant in Essen or Brasilia, and
that except for superficial details the upper-middle-class suburban landscapes
of Jardin in São Paulo resemble those of suburban Sydney or San Francisco.

In terms of geographical space, the expansion of capital and the equalization
of conditions and levels of development are what lead to the so-called "shrinking world," or to "space–time compression" (Harvey 1989). Capital drives to
overcome all spatial barriers to expansion and to measure spatial distance
by transportation and communication time. This is the process which Marx
perceptively labeled the "annihilation of space with time":

> Capital by its nature drives beyond every spatial barrier. Thus the
> creation of the physical conditions of exchange – of the means of
> communication and transport – the annihilation of space by time –

becomes an extraordinary necessity for it. . . . Thus, while capital must on the one side strive to tear down every spatial barrier to intercourse, i.e., to exchange, and conquer the whole earth for its market, it strives on the other side to annihilate this space with time, i.e., to reduce to a minimum the time spent in motion from one place to another. The more developed the capital, therefore, the more extensive the market over which it circulates, which forms the spatial orbit of its circulation, the more does it strive simultaneously for an even greater extension of the market and for greater annihilation of space by time. . . . There appears here the universalizing tendency of capital, which distinguishes it from all previous stages of production.

(Marx 1973 edn.: 524, 539–40)

The economic correlate of this universalizing process is the tendency toward an equalization in the rate of profit (Marx 1967 edn.: III, Ch. 10). Both tendencies are realized in the circulation of capital but express a deeper process rooted in production: the universalization of abstract labor and the consequent hegemony of "value" over social interchange (Harvey 1982; Sohn-Rethel 1978).

To those who have followed the development of urban theory in recent decades, this equalization tendency as it operates at the urban scale will have a familiar ring. But before examining the urban scale *per se*, it is necessary to look at the corollary process of differentiation. The differentiation of levels and conditions of development does not emanate from a single focus but occurs along a number of axes. In the first place, contemporary capitalism clearly inherits an environment that is differentiated according to natural features. This natural basis of differentiation was a fundamental ingredient, in earlier societies, of the uneven *societal* development that occurred. To cite but one example, there developed regional divisions of labor, based on the differential availability of natural materials: textiles where sheep could graze and water power was available, iron and steel where coal and iron ore were available, towns at port sites, and so on. This, of course, was the bread and butter of traditional commercial and regional geography and in part the basis of the descriptive "areal differentiation" tradition in geographic research. But the advanced development of capitalism has brought about a certain emancipation from nature and natural constraints. "Important as the natural differences in the conditions of production may be," wrote Nikolai Bukharin, "they recede more and more into the background compared with differences that are the outcome of the uneven development of productive forces" (1972 edn.: 20). Thus contemporary geographical differentiation, while retaining deeply interwoven remnants from earlier nature-based patterns of differentiation, is increasingly driven forward by a quintessentially social dynamic emanating from the structure of capitalism.

This dynamic involves the progressive division of labor at various scales, the spatial centralization of capital in some places at the expense of others, the evolution of a spatially differentiated pattern of wage rates, the development of a ground rent surface that is markedly uneven over space, class differences, and so forth. It would be a mammoth task to attempt a general dissection of the intricacies of each of these processes and relationships that contribute to the tendency toward geographical differentiation. In any case, these processes and relationships take on a radically different significance depending upon the scale being considered. Wage rates, for example, are one of the central determinants of uneven development at the international and regional scales, but at the urban scale, I would argue, are relatively unimportant. Elaborating the general dynamic of differentiation remains one of the most challenging obstacles to the construction of a general theory of uneven development and will not be pursued further here. Instead, we shall confine ourselves to a discussion of the urban scale where the analysis of differentiation can be made concrete. The essential point at this stage, however, is that a tendency or series of tendencies operate in opposition to the equalization of conditions and levels of development in a capitalist economy, and it is the contradictions between these as they play themselves out in concrete history, that lie behind extant patterns of uneven development. More than anything else, this process of differentiation, counterposed as it is by equalization, is responsible for the opposition of developed versus underdeveloped regions and nations and for the opposition of suburb and inner city.

At the height of the optimism of postwar expansion Melvin Webber (1963, 1964a, 1964b) developed the concept of the "urban non-place realm." Webber reasoned that with the development of new technologies, especially in communications and transportation, many of the old forms of social difference and diversity were being broken down. For an increasing number of people, economic and social propinquity had become emancipated from spatial propinquity; with the exception of the poor, he argued, urbanites had freed themselves from the restrictions of territoriality. Webber's notion of a "non-place urban realm" was given a wide and appreciative airing, not just because its optimism and idealism were wonderfully in tune with the times and because it seemed to express the rising liberal vision of the urban planning profession, but also because, however nebulously, it did express a real, concrete tendency in postwar urban development. What Webber captured, albeit often implicitly and at times obliquely, was the tendency toward equalization as it operated at the urban scale. Against this emphasis on equalization, David Harvey emphasized the opposite process, the differentiation of urban space, and stressed the importance of class beneath this differentiation process (Harvey 1973: 309).

In retrospect, it should be clear that both positions express at least a half-truth. The impetus behind a spaceless urbanism is only accelerated by the advent of computerized work, advances in telecommunications, electronic networking, telecommuting. And yet access to these advances is highly uneven, and many people find themselves trapped in urban space rather than freed from it. Beneath the apparent theoretical contradiction between the "non-place urban realm" and the redifferentiation of urban space therefore lies a real contradiction in the spatial constitution of capitalism.

At the urban scale, the main pattern of uneven development lies in the relation between the suburbs and the inner city. The crucial economic force mediating this relation, at the urban scale, is ground rent. It is the equalization and differentiation of ground rent levels between different places in the metro-politan region that most determines the unevenness of development. In making this assertion, I am aware that other social and economic forces are involved, but many of these are expressed in the ground rent structure. Wage and income levels are certainly expressed in class and race segregation in a city's housing market, but these differences are mediated through ground rent. Or the transportation system, for example, makes some locations more accessible and therefore (generally) more favorable, leading to higher land prices which represent nothing but more highly capitalized ground rent. But there is an obvious chicken-and-egg question here: does a new transportation system restructure the ground rent surface, hence leading to new develop-ment, thus necessitating new transportation systems? Certainly at the urban scale, the latter is the norm where fundamental alterations are concerned. This is the difference between suburbanization, a fundamental process in urban development, and ribbon development, which is relatively ephemeral; although clearly enhanced and encouraged by the development of the means of transportation, suburbanization was a product of deeper and earlier forces (Walker 1978, 1981). Ribbon development, on the other hand, is precisely the case where new transportation routes alter the pattern of accessibility and hence the local ground rent structure, leading to new development that clings exclusively to the new route. Without the new road, railway or canal, development would not have occurred.

The pattern of ground rents in an urban area is highly functional in that it is the mechanism by which different activities are allocated through the land market to different spaces. While managing or mediating this differentia-tion or urban space, ground rent is not in itself the origin of differentiation. Rather, the ground rent surface translates into a quantitative measure of the actual forces tending toward differentiation in the urban landscape. These differentiating forces are of two major sources in the contemporary city. The first is functional in the more specific sense, referring to the difference between residential, industrial, recreational, commercial, transportational and institutional land uses. Within each of these categories there is a differen-tiation according to scale; large-scale modern industrial plants tend to be geographically differentiated from small-scale, labor-intensive workshops, for example. The second force – and this applies mainly to residential land use – is differentiation according to class and race (Harvey 1975). These two sources of social and functional differentiation are translated into a geographical differentiation mainly through the ground rent structure.

But what about wage differentials and the uneven development of urban space? It is often assumed that there is no significant pattern to wage differen-tials across urban space. In an insightful study of Toronto, however, Allen Scott (1981) has detected a distinct and systematic spatial pattern of wage differentials. The farther one goes toward the urban fringe from the core, the higher are the wages. Interpreting this result, Scott suggests that while a number of other factors are important, the higher wages in the suburbs are predominantly the result of the local relationship between supply and demand; where the supply of labor is least, owing to lower densities, namely

the suburbs, wages will be higher, and vice versa. It make sense to see differential wage rates as the result of the suburbanization of industrial and other
employment rather than its cause, since no matter how capital-intensive
suburban fringe companies are, they will move despite, not because of, higher
wages. In fact there is another possible interpretation of the data, suggesting a
more direct relationship between the type and scale of industry and the wage
rate. It is possible that the higher wage rates toward the suburbs are due to the
fact that industries locating in the suburbs tend, on average, to represent
newer, larger, more capital-intensive, more advanced sectors of the economy
where levels of skill, and hence wage rates, are comparatively higher.

The actual history of suburbanization supports treating wage rates more as
the dependent than the independent variable – dependent less on intraurban
population density and more on the nature of the work process. This conclusion would apply for the urban scale only; at the regional and international
scales the opposite pertains (Mandel 1976; Massey 1978).

The urban labor market (unlike the housing market) is not acutely subdivided as a result of direct spatial constraints on access. Essentially, it is a
single *geographical* labor market no matter how differentiated it may be
socially, according to skills and race, class and gender. The urban scale as a
distinct spatial scale is defined in practice in terms of the reproduction of labor
power and the journey to work. The entire urban area is relatively accessible
for most commuters; one can get from city to suburb and from suburb to city
relatively quickly, and with a little more difficulty from suburb to suburb.
Whether or not we accept Scott's explanation of the wage differentials across
urban space, the essential point here is that present patterns of industrial
location at the urban scale are not a product of whatever wage differentials
do exist, but rather, help to create such differentials.

To the extent that the urban area is a single geographical labor market, and
that the developments of the transportation network have extended significantly the area over which the daily commute can be made, the tendency
toward equalization has been realized at the urban scale. But this is equalization in a rather trivial sense. A far more fundamental equalization takes
place historically in the ground rent structure. The traditional ground rent
surface assumed in neoclassical models is usually described as a function or
curve which declines with increasing distances from the center. This surface
is purported to evolve because of the participation of different kinds of actors
in the land market, each with different preferences for space and therefore with
different "bid–rent curves." Thus, when we disaggregate, we get the familiar
result of intersecting curves, each representing a land use with a different
rate of change. If we now disaggregate within residential land uses according
to class, we get the equally familiar result of intersecting income curves:
low income at the center, high income at the periphery. These ideal models of
the urban land market are entirely consistent with the filtering model discussed
in Chapter 3, and while they may have had some empirical validity in earlier
years, they no longer describe the urban ground surface today. Today's
rent gradient more resembles that presented in the weakly bimodal curve in
Figure 3.1.

Plate 4.2 The Eton Center in downtown Toronto

This pattern suggests the operation of both an equalization process and a differentiation process. On the one hand, the development of the suburbs has significantly reduced the general differential between central and suburban ground rent levels for any given location in the suburbs. But on the other hand, a "land value valley" has emerged in the inner city surrounding the central area. This area has been spatially differentiated from surrounding areas, giving it a ground rent level quite at variance with the assumptions implied in the earlier neoclassical bid–rent models. With a different ground rent level, the potential uses of this land are also quite different from those that would be consistent with the neoclassical model.

In order to understand the specific origins of this pattern and to assess the potential for future land uses, it is necessary to make a more historical argument concerning uneven development. This brings us to the second aspect of uneven development to be considered, the valorization and devalorization of capital invested in the built environment.

VALORIZATION AND DEVALORIZATION OF CAPITAL IN THE BUILT ENVIRONMENT

Capital invested in the built environment has a number of special features, but the emphasis here is upon its long turnover period. Whether it is fixed capital invested in the direct production process or capital invested to provide the means or reproduction (houses, parks, schools, etc.) or the means of circulation (banks, offices, retail facilities, etc.), capital invested in the built environment is immobilized for a long period in a specific material form. The valorization of capital in the built environment – its investment in search of surplus value or profit – is necessarily matched by its devalorization. During the period of its use through immobilization in the landscape, the valorized capital returns its value piece by piece. The invested capital is devalorized as the investor receives returns on the investment piecemeal. The physical structure must remain in use and cannot be demolished, without sustaining a loss, until the invested capital has returned its value. This has the effect of tying up whole sections of land over a relatively long period in one specific land use and thereby creates significant barriers to capital mobility and new development. But new development must proceed if accumulation is to occur. In addition to creating barriers to the further valorization of capital in the built environment, therefore, the steady devalorization of capital creates the possibility of its opposites, namely longer-term possibilities for a new phase of valorization through investment, and this is exactly what has happened in the inner city.

Concerning capital invested in housing, the economic devalorization process is often marked by an obvious sequence of transitions in the tenure arrangements, occupancy and physical condition of properties in a neighborhood. This is the downward sequence described in Chapter 3 as the devalorization cycle. This economic decline of inner-city neighborhoods is a "rational," predictable outcome of the free enterprise land and housing markets (Bradford and Rubinowitz 1975; Lowry 1960). Just as the devalorization of capital is implied in its valorization, the decline of the inner cities is implied in the more general expansion of urban areas, and particularly in the development of the suburbs.

As Walker (1981) points out, a number of very complex forces are involved in the development of the suburbs, but it is vital to see suburbanization as complementary to inner-city decline in a wider pattern of uneven development at the urban scale. Suburbanization is the product of the interplay of the processes of equalization and differentiation at the urban scale. Fundamentally, it represents a considerable historical emancipation of urban social form from space. This process has several dimensions. The emancipation of social capital from spatial constraint is part of the more general project of emancipation from but immersion in nature, represented by suburbanization: that is, capital accumulation and expansion and the annihilation of space by time at the urban scale take a quite specific form. An expanding area of the nonurban periphery is brought into the sphere of urban space. In its spatial aspect, this explosive expansion of urban space has been led by the process of suburbanization. In that it progressively reduces all society to urban society, this urbanization of the countryside represents one of the most acute forms of the equalization of conditions of development under advanced capitalism.

"Accumulation of capital . . . is increase of the proletariat," Marx (1967 edn.: I, 614) argued, and indeed the accumulation of capital brings perforce the accumulation of a growing labor force. With the increased social centralization of capital along with the operation of agglomeration economies, there is a strong tendency for new and expanding productive activity to locate itself in urban areas. The social centralization of capital – its concentration in larger and larger quantities in fewer and fewer enterprises – is a direct expression of the constant drive to accumulate (Marx 1967 edn.: I, 625–628), and this social centralization translates in part into a spatial centralization of capital. If this helps explain the explosive urban expansion of the nineteenth and twentieth centuries, there still remains to be explained the differentiation between suburb and inner city. This differentiation was both the product of expansion and the means by which expansion occurred.

The earliest development of upper-class residential suburbs – often seasonal country houses at first – was the spatial expression of two intertwined divisions of labor. In the first place, it represented a gendered division between work and home, or rather, it came to represent this division as the middle class suburbanized, because many of the first elite suburbanites were not employed. But second, it also represented a spatial division between classes insofar as the early suburbs separated the upper and upper middle classes from the urban rabble. Only later did middle- and working-class suburbs emerge, first in Europe and later in North America, where suburbanization also marked a spatial differentiation according to race. Working-class suburbanization followed the suburbanization of industry, which was also, in part, a product of the progressive division of labor, particularly at the scale of the individual plant. As many labor processes were broken down into a larger number of simpler, less skilled tasks, the recombination of these separate activities into a single composite production process required more space. This was partly due to the multiplication in the number of individual tasks, partly due to the increased scale of machines, and partly due to the fact that, to remain competitive, productive units had to be larger. Thus, the division of labor and the necessary recombination of these divisions necessitated an expansion in the spatial scale of the production process. Movement to the suburbs, where ground rent was low, was the only economical alternative. It is not that suburbanization was the only alternative *per se*, it is just that the redevelopment of the established city was not an economical option. The center was still functional, meaning that it was still in the process of devalorization. The compromise between competing impulses for equalization and differentiation, therefore, came with the suburbanization of the urban periphery.

The development of the suburbs should be seen not so much as a decentralization process, more as a continuation of the vigorous centralization of capital into urban areas. Yet, simultaneously, suburbanization enhances the internal differentiation of urban space. Thus the suburbanization of capital from the nineteenth century onward was simultaneously the economic abandonment of the inner city in terms of both new construction and repairs. This process was most acute in the US where state regulations did least to modulate capital mobility, but it is general to the economies of Europe, Australia and North America. It is this spatial shift of capital investment, of course, that led to the rent gap.

The investment of capital in the central and inner city, then, caused a physical and economic barrier to further investment in that space. The movement of capital into suburban development led to a systematic devalorization of inner- and central-city capital, and this in turn, with the development of the rent gap, led to the creation of new investment opportunities in the inner city precisely because an effective barrier to new investment had previously operated there. The issue to be examined now is the rhythm and periodicity of these movements of capital, and this is the third and final aspect of uneven development to be considered.

Reinvestment and the rhythm of unevenness

The rhythm and periodicity in an urban economy is closely related to the broader rhythms and periodicity of the national and international economy. Thus Whitehand (1972) has shown how urban expansion and suburbanization in Glasgow have taken place in consecutive waves occurring at particular points in economic boom–bust cycles (see also Whitehand 1987). As Harvey (1978, 1982) has shown, there is a strong empirical tendency for capital to undergo periodic but relatively rapid and systematic shifts in the location and quantity of capital invested in the built environment. These geographical or locational switches are closely correlated with the timing of crises in the broader economy. Crises are not accidental interruptions in some general economic equilibrium, as neoclassical economic theory would suggest, but are integral instabilities punctuating an economic system based on profit, private property and the wage relation. The necessity to accumulate leads to a falling rate of profit, an overproduction of commodities, and thereby to crisis (Marx 1967 edn.: III, Ch. 13).

Gentrification is intimately intertwined with these larger processes. By way of the simplest explanation, let us begin with falling rates of profit. When rates of profit in the major industrial sectors begin to fall, financial capital seeks an alternative arena for investment, an arena where the profit rate remains comparatively high and where the risk is low. At precisely this point, there tends to be an increase in the capital flowing into the built environment. The result is the familiar property boom, such as affected a number of cities throughout the advanced capitalist world from 1969 to 1973 and in the late 1980s. But the question of *where* this capital flooding into the built environment will locate has no automatic answer. It depends in part on the geographic patterns created in the foregoing economic boom. In the case of the present restructuring of urban space, the geographical pattern confronting capital was created through the simultaneous development of the suburbs and inner-city underdevelopment. The underdevelopment of the previously developed inner city, brought about by systematic disinvestment, provoked a rent gap which, in turn, laid the foundation for a locational switch by significant quantities of capital invested in the built environment. Gentrification in the residential sphere is therefore simultaneous with a sectoral switch in capital investments.

This locational switch is rarely smooth, as is illustrated by the dramatic fluctuations in new housing construction that have accompanied the booms and busts of most national economies since the 1970s. Uneven development at the urban scale therefore brought not only gentrification in the narrowest sense but the whole gamut of restructurings: condominium conversions, office construction, recreational and service expansion, massive redevelopment projects to build hotels, plazas, restaurants, marinas, tourist arcades, and so on. All involve a movement of capital not simply into the built environment in general, in response very much to the approaching or already present economic crisis, but into the central and inner urban built environment in particular. The reason for this particular geographical focus of reinvestment can be found in the historical patterns of investment and disinvestment that represented the inner city as an opportunity for reinvestment. (In this light, incidentally, it makes sense to reassess the traditional liberal view that the 1950s state-subsidized urban renewal schemes in the US were a failure. Regardless of how *socially* destructive urban renewal was – and it was socially destructive – it was actually very successful *economically* in laying the foundation for the phase of redevelopment, rehabilitation, land use conversion and, ultimately, private-market gentrification that would follow (Sanders 1980)).

Economic crisis both necessitates and provides the opportunity for a fundamental restructuring of social and economic space. In the US, suburbanization was a concrete spatial response to the depressions of the 1890s and 1930s, in the sense that suburban development opened up a whole series of investment possibilities which could help to revive the profit rate. With FHA mortgage subsidies, the construction of highways, and so on, the state subsidized suburbanization quite deliberately as part of a larger solution to crisis (Walker 1977; Checkoway 1980). Albeit a reversal in geographic terms, the gentrification and redevelopment of the inner city represents a clear continuation of the forces and relations that led to suburbanization. Like suburbanization, the redevelopment and rehabilitation of the central and inner cities functions as a substantial engine of profit.

Gentrification is part of the restructuring of inner-city residential space. It is integral with the preexisting restructuring of office, commercial and recreational space, and, while this restructuring has a variety of functions, it operates primarily to counteract the falling rate of profit. In his National Urban Program, President Jimmy Carter implicitly understood this. For the first time, the "revitalization of the cities" was seen as integral to the overall revitalization of the US economy. This implicit realization was symbolized by Carter's attempt to create a new government department by consolidating the Department of Housing and Urban Development (HUD) and the Economic Development Administration into the Department of Development Assistance. The program never came to fruition, of course, but it was an ambitious state plan to lubricate the restructuring of urban space in the name of national economic revitalization. Since 1980, the governments of the US, the UK and most other advanced capitalist economies have taken a quite different tack, effectively withdrawing or circumscribing state involvement in housing investment. Gentrification has of course thrived in this new climate of privatization.

While gentrification represents the leading edge of spatial restructuring at the urban scale, deindustrialization, globalization, resurgent nationalisms, the EU and the newly industrializing countries all signal a spatial restructuring at the global, national and regional scales (Harris 1980b, 1983; Massey 1978; Massey and Meegan 1978). And while the urban scale may in the end be the least significant in terms of the overall restructuring of the world economy, the internal logic of uneven development is most completely accomplished there. The logic of uneven development is that the development of one area creates barriers to further development, thus leading to an underdevelopment that in turn creates opportunities for a new phase of development. Geographically, this leads to the possibility of what we might call a "locational seesaw": the successive development, underdevelopment and redevelopment of given areas as capital jumps from one place to another, then back again, both creating and destroying its own opportunities for development (N. Smith 1984).

There are clearly limits to the possible extent of this locational seesawing. At the international scale, where, with few exceptions, the distinction between developed and underdeveloped nations is rigidly set by national boundaries and military defenses, the process is difficult to see in any accomplished form. At the regional scale, however, some previously developed industrial regions such as New England, central Scotland, northern France and the Ruhr have experienced precipitous decline followed by some reinvestment in the latter half of the twentieth century, suggesting at least a mild version of this seesaw. This seesawing is perhaps most complete at the urban scale, and more complete with US cities than elsewhere. This is the significance of gentrification: once developed, then underdeveloped, the central and inner cities are again in the midst of an active redevelopment.

It is important to stress that this does not imply an imminent end to suburbanization; just as new construction and repairs continued in the city during the most vigorous period of suburbanization, the urbanization of the countryside will also continue, with the emphasis increasingly on the areas beyond the present suburbs (Garreau 1991). This is clear, if for no other reason than that central and inner-city redevelopment, while it can absorb massive quantities of capital in the process of economic restructuring, can never be the exclusive geographical focus for reinvestment. The scale at which economic restructuring is necessary will ensure that central- and inner-city redevelopment from only a small part of the overall restructuring process. The differentiation of the city from the suburbs, through redevelopment and the probable decline and underdevelopment of selected suburbs, will be matched by the continued urbanization of the countryside.

CONCLUSION

In the introduction to this work, I suggested that revitalization was rarely an appropriate term for gentrification, but we can see now that in one sense it is appropriate. Gentrification is part of a larger redevelopment process dedicated to the revitalization of the profit rate. In the process, many downtowns are being converted into bourgeois playgrounds replete with quaint markets, restored town houses, boutique rows, yachting marinas and Hyatt Regencies. These very visual alterations to the urban landscape are not at all an accidental side-effect of temporary economic disequilibrium but are rooted in the structure of capitalist society every bit as deeply as is suburbanization. The economic, demographic, lifestyle and energy factors cited by exceptionalists and optimists alike are relevant only after consideration of this basic explanation in terms of uneven development at the urban scale. Several studies tend to confirm the somewhat countercyclical nature of gentrification *vis-à-vis* long swings in the economy and economic crisis prior to the late 1980s, whether in Atlanta and Washington DC (James 1977: 169) or in Canadian cities (Ley 1992), although Badcock's (1989, 1993) work in Adelaide presents a more mixed picture. The picture also changed with the depression of the late 1980s and early 1990s, but that is the subject for later discussion (Chapter 10).

Gentrification, and the redevelopment process of which it is a part, is a systematic occurrence of late-capitalist urban development. Much as capitalism strives toward the annihilation of space by time, it also strives more and more to produce a differentiated space as a means to its own survival. A predictably populist symbolism underlies the hoopla and boosterism with which gentrification is marketed. It focuses on "making cities livable," meaning livable for the middle class. In fact, and of necessity, they have always been "livable" for the working class. The so-called renaissance is advertised and sold as bringing benefits to everyone regardless of class, but available evidence suggests otherwise. For instance, according to the Annual Housing Survey conducted by the US Department of Housing and Urban Development, approximately 500,000 US households are displaced each year (Sumka 1979), which may amount to as many as 2 million people. Eighty-six percent of those households are displaced by private-market activity, and they are predominately urban working class. Even while liberal urban policy survived in the 1970s, the federal government sidestepped the problem of displacees, claiming alternately that there are no accurate data on displacement, that it is an insignificant process compared to continuing suburbanization, or that it is the responsibility of local government (Hartman 1979). Further, the so-called renaissance is generally sold as a means to raise the cities' property tax revenues and decrease unemployment, but there is little evidence that these benefits have occurred either. Not until the surge of homelessness in the 1980s made the connection between gentrification and its costs explicit did national and local governments in the US begin to deal even summarily with the social fall-out from the restructuring of urban space. Part of the federal government's response was to stop collecting the kinds of figures quoted above.

Since the 1970s, the economic restructuring that succeeded the postwar political economy has reached into every corner of economic and social activity. Through gentrification as well as service cuts, unemployment and attacks on welfare, the restructuring of working-class communities – the reproduction of labor power – is itself attacked as part of this larger economic restructuring. During the 1970s, and perhaps again in the wake of the 1980s, it is becoming increasingly clear that the struggle over the use and production of space is heavily inscribed by social class (as the nomenclature of "gentrification" itself suggests) and race as well as gender. Gentrification is thereby part of the social agenda of a larger restructuring of the economy. Just as economic restructuring at other scales (in the form of plant closures, runaway shops, social service cuts, etc.) is carried out to the detriment of the working class, so too is the spatial aspect of restructuring at the urban scale: gentrification and redevelopment.

Plate 4.3 Global graffiti against gentrification: (clockwise from top) Sydney, New York, Goslar (Germany), Granada (Spain)

5

SOCIAL ARGUMENTS

Of yuppies and housing

Nineteen eighty-four was not the year of George Orwell, according to *Newsweek*, but rather "The Year of the Yuppie." Or so read the front cover of the year-end issue, and it is no accident that the first photograph in the accompanying article identified the yuppie lifestyle with gentrification. Coined apparently in 1983 to refer to those young, upwardly mobile professionals of the baby-boom generation, the term "yuppie" has already achieved a wide currency; few words have had such an impressive debut in the language. Apart from age, upward mobility and an urban domicile, yuppies are supposed to be distinguished by a lifestyle devoted to inveterate consumption. To the popular press, therefore, which generally extols the virtues of gentrifying urban "pioneers," the link between the two icons – yuppies and gentrification – was irresistible. In the academic literature, traditional explanations have also emphasized the role of consumption choices, lifestyle changes and the baby boom generation, but a number of researchers seeking more rounded explanations have begun to conceive gentrification as the social and geographical correlate of the rise of the yuppie, or, in more sober terms, the development of a "new middle class." More generally, gentrification is treated as the result of a contemporary social restructuring (Mullins 1982; Rose 1984; Williams 1984a, 1986).

In the preceding chapters, I have treated gentrification as a product of political economic shifts in local and global markets. The purpose of this chapter is to examine the more social dimensions of gentrification, and specifically the claim that gentrification results from the social restructuring that affected many national societies beginning in the 1970s. I begin with statistical evidence on the existence of the new middle class, then consider the claim that the changing social roles of women constitute a significant impetus to gentrification. Gentrification is not a "chaotic concept," as some have claimed, but the class and gender restructurings that accompany it are not simple either. To what extent does an understanding of class and gender in the urban landscape help to connect economic and social visions of gentrification?

A NEW MIDDLE CLASS?

Who comprises this new middle class? In a vivid portrait, Raphael Samuel argues that the new middle class

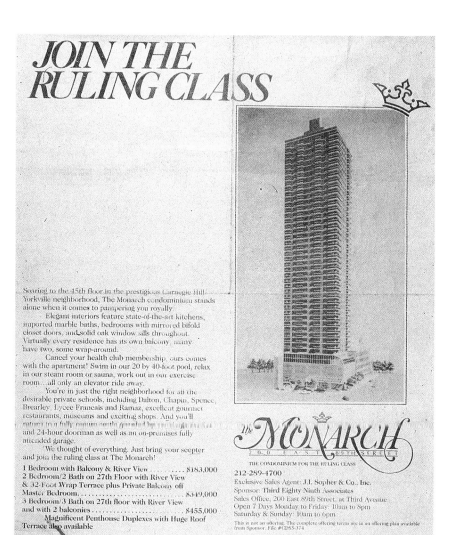

Plate 5.1 "Join the ruling class": newspaper advert for a new condominium in New York City

distinguishes itself more by its spending than its saving. The Sunday colour supplements give it both a fantasy life and a set of cultural cues. Much of its claim to culture rests on its conspicuous display of good taste, whether in the form of kitchenware, "continental" food, or weekend sailing and cottages. New forms of sociability, like parties and "affairs," have broken down the sexual apartheid which kept men and women in rigidly separate spheres. . . .

The new middle class are outward looking rather than inward looking. They have opened up their homes to visitors, and exposed them to the public gaze. They have removed the net curtains from their windows, and taken the shutters down from their shops. They work in open-plan offices and establishments, with plate-glass windows and see-through partitions and doors. In their houses they make a fetish of light and space, replacing rooms with open-access living areas and exposing the dark corners to view. . . . Class hardly enters into the new middle class conception of themselves. Many of them work in an institutional world of fine gradations but no clear lines of antagonism. . . .

The new middle class have a different emotional economy from that of their pre-war predecessors. They go in for instant rather than deferred gratification, making a positive virtue of their expenditure, and treating the self-indulgent as an ostentatious display of good taste. Sensual pleasures, so far from being outlawed, are the very field on which social claims are established and sexual identities confirmed. Food, in particular, a postwar bourgeois passion . . . has emerged as a critical marker of class.

(Samuel 1982: 124–125)

There are obvious national features in this British portrait but its underlying flavor is instantly recognizable in a variety of national contexts. But far from being a recent phenomenon, as the popular perception holds, discussion of the "new middle class" can be traced to the turn of the twentieth century. According to the historian Robert Wiebe (1967: 111–132), talking in the US context, this group of urban professionals, experts and managers experienced a "revolution in identity" as the specialized needs of the emerging urban industrial system gave them an increasingly prominent social role. Individuals in this "new middle class," he says, were imbued with a "confident driving quality" and harbored "an earnest desire to remake the world upon their private models."

But once we move forward from these "proto-yuppies" of the Progressive Era, agreement is overtaken by ambiguity. Despite decades of debate there is not only no generally accepted definition but not even any agreement on what we might call a general definitional arena for this new middle class. This same social group is conceptualized in a variety of specific niches on the social totem pole, and the array of different labels attached to them testifies to this. Apart from the "new middle class" and the professional-managerial class, the social science literature is replete with concepts of a "new class" (Bruce-Briggs 1979), a "new working class" (Miller 1965), a "salaried middle class" (Gould 1981), "middle strata" (Aronowitz 1979), a "working middle class" (Zussman

1984), a "professional middle class" (Ehrenreich and Ehrenreich 1979), and so on – not to mention the simple staid "middle class" of old. In short, while these different concepts intersect, more or less, on the class map, it is not quite clear to whom they refer. Indeed the notion of class itself is subject to many interpretations. For my purposes here, I will take as axiomatic the broad proposition that class is defined according to people's social relations to the means of production.

The question of what makes it a *new* middle class is particularly important here. In an article that spurred some of the resurgence of interest in the topic in the last fifteen years, the Ehrenreichs (1979) argued that unlike the old middle class of artisans, shopkeepers, independent farmers and self-employed professionals, the professional-managerial class was not independent of the capital–labor relation but was employed by capital for the purposes of managing, controlling or administering to the working class. It now constitutes an estimated 25 percent of the US population. Others less analytical have tended to treat this group as synonymous with the entire "white-collar" labor force, which could put the new middle class at closer to 60 percent of the population. Nicos Poulantzas, in his structuralist scaffolding of the contemporary class system, wedged the new middle class in between the working class and the capitalist class as that group of functionaries who neither own the means of production nor perform productive labor but who are political and ideological participants in the domination of the working class (Poulantzas 1975). In a more sophisticated analysis, Erik Olin Wright (1978) has rejected attempts to straitjacket society into different class corners, and insists instead that we have to recognize the reality of "contradictory class positions"; classes more resemble fuzzy sets than discrete pigeonholes. The new middle class, for Wright, is the classic example of contradictory class location. This group is pulled hither and thither by the economic aspirations of the class above them, the political potential of the class below them and the ideological dictates of their daily occupations. More traditional analyses, many of them dating from the 1950s, attempt to define this new class on the basis of consumption patterns, providing a consumption-side corollary for the "white-collar" argument (see Parker 1972).

The new middle class also has a highly ambiguous political profile and this would tend to lend support to Wright's notion of contradictory class locations. Wiebe's proto-yuppies, as I have called them, were clearly "progressives" (in the sense of the Progressive Era in the US), and the Ehrenreichs (1979) depict the leaders of the New Left as coming from the same (if expanded) class six decades later. Indeed the term yuppie emerged in the US in 1983 in connection with the candidature of Gary Hart for the presidency, and although neither he nor his supporters had roots in the New Left, they might be characterized as latter-day conservative progressives, the "neos" of the 1980s – neoliberal and/or neoconservative. By the Clinton presidency of the 1990s, the administration and its opponents alike were dominated by the neos. In Australia, the political stereotype of new middle-class "trendies" is that they are personally conservative but socially conscious activists in the Australian Labor Party. In the British context, the new middle class was conceived as spanning the spectrum from "trendy lefties," through the core supporters of the more neoliberal Social Democratic Party (SDP), now Liberal Democrats, to the young conservatives. In the words of a columnist for the *Financial Times* of London, the formation of the SDP represented "primarily a sociological development, an example of the political system beginning to catch up with societal change. . . . There is a new class which outnumbers either the stereotypes of working class or capitalists" (Rutherford 1981).

Ambiguous as the notion of a new middle class might be, some commonly accepted themes are implied, and this allows us to go some way toward identifying the class in practice. Given the ambiguity of definitions, however, there are serious problems in translating between theory and empirical identification; the two avenues of identification discussed here – structural and economic – should therefore be seen not as definitions of class so much as indicators of class specificity. In the first place, the new middle class is deemed to be the product of an altered occupational and income structure; the pattern of change is very familiar. The Western capitalist economies have experienced a decline in the relative importance of manufacturing employment and a parallel increase in the importance of professional, administrative, service and managerial occupations, especially in producer services (finance, insurance, real estate and such), nonprofit services (mainly health and education) and the government sector. In Britain, for example, the percentage of employees classed as professional rose from 6.6 percent in 1951 to 19.1 percent in 1993 while the percentage of agricultural, industrial and other manual employees fell from 72.2 percent to 49.3 percent during the same four decades (Routh 1980; International Labour Organization 1994).

The transformation in occupational structures is undeniable, but we should not jump too quickly to the conclusion that this is tantamount to the emergence of a new middle class: class divisions cannot be uncritically equated to occupational differences. As regards gentrification it is also undeniable that professional, managerial and upper-level administrative personnel in the expanding sectors are heavily represented among gentrifiers: a host of survey-centered case studies have established this statistical generality (Laska and Spain 1980). But it may not be the transformation of the occupational structure that is most crucial to this link between yuppies and gentrification.

The claim that gentrification emanates from contemporary patterns of social restructuring carries with it the implication not only that employment structures have changed, but that this new middle class is also distinguished economically by disproportionate wealth. The patterns of consumption associated with the new middle class, including patterns of housing consumption, are presumed to result from the higher incomes and the greater spending power of this group. In short, we would expect that the emergence of a new middle class would result in an increase in the aggregate share of income earned by this social stratum – an identifiable redistribution of income toward the centre. The ideology of the new middle class, after all, includes its tales of latter-day Horatio Algers who made it from the slums to Wall Street or the City of London. That is why they are "young *upwardly mobile* professionals." Thus although income differentiation too is in no way synonymous with class differentiation, in the specific argument linking gentrification and the new middle class it would be expected that a relative increase in income share would characterize the rise of this class.

But when we examine income distribution over the past several decades, the pattern is not so simple. Far from suggesting a redistribution of income, the aggregate data present a picture of remarkable stability overlain with cyclical fluctuation. Despite postwar economic growth, the poorest 20 percent of the US population did not earn a significantly greater proportion of the social pie and nor did the richest 20 percent have to relinquish its half of the pie (Table 5.1). If there is any fluctuation from this stable distribution of income, it suggests rather that the minimal democratization of incomes that pertained into the mid-1970s was significantly reversed by the 1980s. By the 1990s, the disparity between rich and poor was greater than at any time in the last quarter-century. As regards a new middle class, presumably located in the third and fourth quintiles, their numbers remained very stable through the 1970s but actually fell significantly beginning in 1982. Far from suggesting the rise of a new middle class, the 1980s, which witnessed the most intense gentrification, would seem to have corresponded with an actual shrinking of the new middle class. If the polarization in terms of income and wealth since the late 1970s is more extreme in the US, a parallel if more muted shift in income distribution has occurred elsewhere in the advanced capitalist world, linked to broader economic shifts.

Table 5.1 Share of aggregate household income in the US, 1967–1992

| Year | Distribution of income | | | | | |
	Lowest fifth	Second fifth	Third fifth	Fourth fifth	Highest fifth	Top 5%
1967	4.0	10.8	17.3	24.2	43.8	17.5
1970	4.1	10.8	17.4	24.5	43.3	16.6
1975	4.3	10.4	17.0	24.7	43.6	16.6
1980	4.2	10.2	16.8	24.8	44.1	16.5
1985	3.9	9.8	16.2	24.4	45.6	17.6
1990	3.9	9.6	15.9	24.0	46.6	18.6
1992	3.8	9.4	15.8	24.2	46.9	18.6

Source: US Department of Commerce, Bureau of the Census 1993. *Money Income of Households, Families, and Persons in the United States: 1992.* Series P60–184.

It is difficult to avoid the conclusion that the appeal of the idea of a "new middle class" – at least in the context of gentrification – is as an "empirical generalization" (Chouinard *et al.* 1984) rather than a theoretical category. It provides a deceptively neat characterization of a process that seems intuitively obvious to most of us, but which really we do not understand. To put it this way is to suggest that while something resembling a new middle class may have emerged, it may not have been very significant, at least in relative economic terms. There are several possibilities:

(a)   The new middle class has a clear economic and occupational identity but its significance is exaggerated in such highly visible experiences as gentrification.
(b)   The new middle class does not distinguish itself on the basis of income so much as occupational, political or perhaps cultural criteria. Professional, managerial and administrative work presumably engenders a distinct self-conception of one's social role and this may translate into equally distinct consumption choices resulting in a spatial concentration in the central and inner city. Absolute increase in income for those joining the class makes this spatial concentration possible.
(c)   The new middle class is not a distinct group by any criteria and the explanation for gentrification must be sought elsewhere.

Before we explore the viability of these alternatives, it is worth expanding the question of the social contours of gentrification. For the argument connecting social restructuring to gentrification refers not only to issues of class constitution but to gender: the changing roles of women (and men), the contemporary transformation of reproduction practices, and the changing relationship between waged and salaried work and reproduction.

WOMEN AND GENTRIFICATION

As Rose (1984: 62) has observed, "it is now increasingly accepted that women are playing an active and important role in bringing about gentrification." The reasons for this participation "have not yet been adequately conceptualized," she says but suggests that larger and larger numbers of women may be led to gentrification either because they can afford such housing for the first time or because they cannot afford anything else. The general case for the importance of women in gentrification is perhaps most succinctly put by Ann Markusen:

> gentrification is in large part a result of the breakdown of the patriarchal household. Households of gay people, singles, and professional couples with central business district jobs increasingly find central locations attractive. . . . Gentrification in large part corresponds to the two-income (or more) professional household that requires both a relatively central urban location to minimize journey-to-work costs of several wage earners and a location that enhances efficiency in household production (stores are nearer) and in the substitution of market-produced commodities (laundries, restaurants, child care) for household production.
>
> (Markusen 1981: 32)

This proposition linking women and gentrification has remained a general affirmation with little documentation of empirical trends. This in itself is rather odd given the extent to which gentrification research has been dominated by an empiricist tradition, yet few of the myriad case studies and neighborhood surveys have involved explicit documentation of the extent of women's involvement in gentrification (but see Rothenberg 1995). What I want to do briefly in this section is to provide some statistical support for this link, from both national (US) and local data.

It is a well-known fact that there has been a virtually steady increase in women's labor force participation since World War II. From 30.8 percent in the US in 1946, the figure has risen steadily to 57.9 percent in 1993 (US Department of Commerce 1994). (For men, by comparison, the rate dropped from 83 percent to 78 percent.) Increased participation in the "official" labor market also seems to have been matched by higher relative incomes. Whereas the ratio of women's to men's median income in 1970 was a mere 33.5 percent, that figure increased steadily to 52.2 percent by 1992 (US Department of Commerce 1994). But this increase was very unevenly distributed, with the greatest proportionate gains going to the highest-income women. Whereas in 1970 only 8.9 percent of working women earned in excess of $25,000, by 1992 the figure had risen to 19.4 percent. Of this group of higher-income women, 87 percent are white. By comparison, the number of men in this income bracket remained quite steady at between 40 percent and 44 percent throughout the period, falling lower during periods of recession as in the early 1980s and 1990s, and rising higher during periods of economic expansion (early 1970s and late 1980s). What this suggests is that in absolute terms, there was indeed an expanding high-income population, and that women, especially white women, were a steadily increasing if always minority proportion of this group. It suggests too that the increase in the number of upwardly mobile women is matched by a relative compensatory drop in the incomes of women at the bottom of the income scale. This of course is consistent with the feminization of poverty argument (Stallard *et al.* 1983; Scott 1984).

The significance of these US data is that at the top end of the income hierarchy a significant expansion in the number of women is taking place and this group does indeed represent a reservoir of potential gentrifiers. They are upwardly mobile and, as Rose suggests, may be in the position of affording relatively salubrious housing for the first time. But how does this aggregate national picture compare with the local situation? In the absence of a comprehensive survey, let me present the picture that emerges from ongoing research work in New York City. Given its specificity, this portrait should be treated as only a partial indicator of the extent of women's involvement in gentrification.

All five neighborhoods examined here experienced gentrification by the 1970s and this carried forward into the 1980s and, to a more moderate extent, the 1990s. Census data on income and rent between 1970 and 1990 confirm the obvious social and physical transformations that have taken place. The neighborhoods are quite varied in their physical and social make-up. The first area is Greenwich Village, which is predominantly residential and, for all its cultural frenzy, comprises an established set of communities. Much of the more recent activity has been on the fringes of the Village, although the West Village, which has attracted predominantly gay gentrification, was already gentrifying by the late 1950s. SoHo and Tribeca are the second and third neighborhoods, lying on the southern border of the Village; they are erstwhile industrial zones dominated by converted warehouses and lofts (Zukin 1982; Jackson 1985), and SoHo is designated an Artists' Zone.

The fourth area is in the Upper West Side immediately adjacent to the Lincoln Center, and while predominantly residential it also includes Columbus Avenue with its upscale restaurants and boutiques. The fifth area is Yorkville, which straddles the border between the Upper East Side and East Harlem; this area has experienced a gradual transformation of the existing housing stock in the 1970s but in the 1980s became the target for an intense program of luxury housing construction (see Figure 7.1, p. 141).

The results of this analysis are striking. In the first place, while the population of New York City declined by more than 10 percent during the 1970s, 75 percent of the tracts examined experienced increases in their overall population, indicating a reconcentration in neighborhoods where abandonment or at least emigration had previously occurred, or else a migration into neighborhoods such as Tribeca which had been dominated by industrial, commercial and other nonresidential land uses. In the 1980s, 67 percent of the tracts exceeded the citywide increase of 4 percent. Even more striking is that, with only two exceptions, the female population in these tracts during the 1970s increased more rapidly than the male; an increasing percentage of residents in these gentrified tracts were female. Further, the two tracts which are exceptions to this rule, and which actually experienced declines in their female population, are in that part of the West Village where gay men have led and dominated the gentrification process. In the 1980s, this disproportionate increase in the number of women in gentrified areas continued, with the increase in the female population outpacing that of men in all but four tracts.

Other than the dramatic increase in their relative numbers, the profile of change among women is akin to what one would expect in gentrifying neighborhoods. They disproportionately comprise single women, especially in the Upper West Side and Yorkville, where the increase in single women living alone or sharing households was matched in the 1970s by an absolute decline in the number of married women as well as female-headed households. The latter probably results from the displacement of poorer families and female-headed households in favor of wealthier single women. As regards age, there was a considerable increase over and above citywide levels of women between the ages of 25 and 44, while the most important change vis-à-vis employment was a relative increase in official employment for those women living in households with husbands present. That is, a higher proportion of families are bringing in two or more salaries. Finally, it appears that an increasing percentage of residents in these neighborhoods have professional, managerial and technical occupations.

Whether these findings are replicated in other cities or whether New York City is in this as in other respects unique remains to be seen. But the evidence from these New York neighborhoods does not seem unexceptional. It lends support to the argument of a link between women and gentrification. More difficult to discern, however, is precisely what role women do play. It would be wrong to conclude that in women we find the premier agency behind gentrification; correlation is not causation, involvement not necessarily instigation. There are really two questions here: first, among women, is it the better economic fortunes of a relatively few women at the top of the income hierarchy that lie behind women's involvement in gentrification – an essentially economic explanation – or is it the political and structural changes in the labor market and in styles and modes of reproduction, provoked by the feminist movement, which have loosened previously oppressive social bonds, albeit again affecting only a specific segment of women defined by class and race? Second, to what extent do women play a specific and different role in gentrification *as* women?

The very language of "gentrification" suggests a class-based analysis of this aspect of urban change, and it is likely that the social explanation of gentrification involves some imbrication of class and gender constitution (Bondi 1991a, 1991b). This does not mean that the economic explanations ought to be abandoned, as Warde wants (1991), reinstating consumption as the prime driving force of urban change (see also Filion 1991). Rather it means that social and economic arguments have to be made to complement each other. In this context I think that Bridge (1994, 1995) is right to insist that social arguments about gentrification consider class constitution in a much wider sense and at wider geographical scales than that of the neighborhood.

IS GENTRIFICATION A CHAOTIC CONCEPT?

If gentrification is a class-rooted process imbued with gender from the start, does this mean, as Rose (1984) has argued, that gentrification is necessarily a "chaotic conception"? Elaborating on a casual comment by Marx (1973 edn.: 100), Andrew Sayer (1982) has proposed that in many of our analyses we employ concepts which are not up to the task we demand of them. In general, these "chaotic concepts" are ill-defined and incapable of grasping the real situations they are meant to convey. But Sayer has a more specific definition in mind concerning the abstraction process through which we derive concepts:

> When we *abstract*, we isolate particular one-sided aspects of objects. We know that these aspects rarely stand in visible isolation spontaneously in the real world, and the purpose of experiments in natural science is precisely to achieve or objectify this abstraction. A "rational" abstraction is one which isolates a significant element of the world which has some unity and autonomous force. On the other hand, a poor abstraction or "chaotic conception" combines the unrelated or divides the indivisible.
>
> (Sayer 1982: 70–71)

The meat of the epistemological problem, of course, according to this perspective, is to distinguish those aspects of reality that are unrelated and those that are indivisible, and to devise concepts accordingly. That is, we seek to develop concepts which bound these "isolated aspects of objects" exactly, including nothing that should be omitted and omitting nothing that should be included.

Rose (1984) applies this epistemological realism to gentrification. It is a complex argument, difficult to disentangle, but worth trying to summarize here. The basic point is that "the terms 'gentrification' and 'gentrifiers'," as commonly used in the literature, are "chaotic conceptions" because they obscure the fact that a multiplicity of processes, rather than a single causal process, are responsible for the transforming inner city. For Rose, gentrification is too narrowly defined in economic terms, and she attempts a conceptual reconstruction. She focuses on changing household structures, alternative lifestyles and the transformation in forms of reproduction of people and labor power; for Rose, then, as for Beauregard (1986) and Williams (1986), the primary question to be confronted is how potential "gentrifiers" come to be produced and reproduced. To explore some of the complexities of the process, she introduces the notion of the "marginal gentrifier," who, she says, may well be female, has only a very moderate income, certainly does not fit our paradigmatic conception of the gentrifier, but may nonetheless play an important part in the gentrification process. She gives the example of a college-educated single parent earning $195 per week in the mid-1980s who inhabited a small studio in the "not-too-nice area of Oakland" (Rose 1984: 67).

I think two arguments are being conflated here, and when we separate them much of the chaos Rose perceives in "gentrification" evaporates. The first argument concerns the increasing difficulty which many poor people have finding reasonable and affordable housing in a "nice area." Women are particularly affected by the problems resulting from lower incomes, and certain groups of women are more affected than others: minority women, single parents, lesbians and unemployed women. Rose's second argument is that as gentrification proceeds, it loses its narrow class character, as "white-collar households of much more modest incomes than the type who gave gentrification its name" become involved. These are two separate arguments and still today they refer to two separate populations – gentrifiers and the poor – yet in her concept of the "marginal gentrifier," Rose obscures the difference between these diverse aspects of urban change and the different populations they affect. There is little doubt that gentrification has become a housing option for a larger reservoir of people in general and women in particular, and in that sense the opportunity to gentrify has filtered down the economic hierarchy and across the political field, but it has hardly filtered down so far that single-parent households earning barely $10,000 per year should be considered gentrifiers. To include such a household under the rubric of "gentrification" is to force a chaos on the term which I do not think it has. It unwittingly aligns Rose with the Real Estate Board's own self-interested redefinition of gentrification (Chapter 2). Further, it is difficult to reconcile Rose's position with the experience in the New York City neighborhoods considered above, where gentrification was accompanied by a decline rather than increase in the number of so-called female-headed households. In those neighborhoods there was a divergence or polarization between rich and poor women, as well as a "marginalization of non-family households" (Watson 1986), not the kind of convergence that is implied in the notion of marginal gentrifiers.

This is not to say that poor women and men, college educated or not, do not at times move into cheap, disinvested areas which might also be experiencing

Plate 5.2   The Porsches of Blythe's Wharf

the beginning of some kind of gentrification. Quite clearly this does happen. The important point here is that gentrification is a process, not a state of existence, and in good realist fashion it ought to be defined at its core rather than its margins. Thus the importance of "marginal gentrifiers" is not that they define gentrification but precisely that they are *marginal* to a process defined as the change "of inner-city neighborhoods from lower to higher income residents" (Rose 1984: 62). Marginal gentrifiers are important, especially in the earlier stages of the process, and may well be distinguished by cultural attributes and alternative lifestyles (Zukin 1982; DeGiovanni 1983), but to the extent that the process continues and property values rise, their ability to remain in the area depends less on their cultural than their economic portfolio. Thus it is the concept of the marginal gentrifier which is, in the end, chaotic. Yet it could be an important notion, carrying considerable descriptive and historical validity; it conveys the hitherto unconceptualized evolution of gentrification from a socially restricted into a somewhat broader process. But it will remain a chaotic conception until it is decoupled from the question of the central defining characteristics of gentrification.

Finally, an implicit political argument lies behind the proposition of marginal gentrifiers as proof of the chaos of "gentrification." Rose argues that since "we cannot put an end to all gentrification," and since there are very different groups involved, all with different interests, broad-based political alliances and coalitions offer the best hope for "progressive types of intervention" and the identity of "oppositional spaces" in which we might experiment with "prefigurative" ways of living and working. We "ought not to assume in advance that all gentrifiers have the same class position as each other and that they are 'structurally' polarized from the displaced" (Rose 1984: 68). Indeed, such a rigid assumption would not tell us much about the gentrification of specific neighborhoods, nor indeed would it especially assist "oppositional practices" and "prefigurative modes of living," but at the same time this should not blind us to the fact that there is a very clear polarization ("structural" or otherwise) between people who participate as gentrifiers and those thereby displaced. One of the consequences of the concept of "marginal gentrifiers" is surely the minimization, in the name of coalition building, of the evident polarization that takes place in many gentrifying neighborhoods.

GENTRIFICATION, CLASS AND GENDER: SOME TENTATIVE CONCLUSIONS

The difficulty in identifying a new middle class, especially in economic terms, should give us pause before we glibly associate yuppies and gentrification. It could well be that "yuppies" and the "new middle class" are merely empirical generalizations (Sayer 1982; Chouinard et al. 1984), providing a deceptively neat characterization of a process that seems intuitively obvious yet unexplained. There is no doubt that employment structure has changed dramatically and that a profound social restructuring is taking place (Mingione 1981), and that it is altering both the traditional roles of women – in and between the home and the workplace – and the class configuration of society. Equally, this social restructuring is heavily implicated in the gentrification process.

Rose is right to reject abstract functionalist (not to be confused with structuralist) treatments of class as a means to comprehend gentrification. Such class analysis marked the early period in the rediscovery of Marx and marxism by social scientists – the effort to begin, however crudely, to see the class character of the societal processes shaping the geographical landscape. The debate has clearly moved beyond this level, which means not that class is irrelevant but rather that class analyses must be more sophisticated. The two-class model of classical marxism does give us considerable insight into capitalism as a whole, but as a tool for comprehending the specific experiences of social and political change it always needs to be refined and developed (cf. Marx 1967 edn.: I 640–648, III 370–371, 814–824; Marx 1963 edn.; 1974 edn.). For examining the details of gentrification in a given neighborhood, it has the effectiveness of a chain saw for wood carving. Not that the two-class model is intrinsically blunt; it is necessary for cutting and shaping the block out of which our more intricate carve of gentrification can be fashioned. Rather, it is a case of misapplication. Just as the two-class model cannot cut the sharp outlines of gentrification, the intricate, more refined and more contingent tools of class analysis appropriate for portraying gentrification in a particular local setting are ineffective for explaining the larger historical and theoretical patterns of capitalist society. They would be as nail files to a forest.

By way of example, Roman Cybriwsky (1978) offers a rich and detailed view of social conflict in a gentrifying neighborhood in Philadelphia, a view which is not at all rooted in a marxist analysis of class, yet one which is not inconsistent with such an analysis. This despite the obvious fact that Cybriwsky relates a sad tale of white alliance, between "gentrifiers" and threatened working-class residents, on the basis of racial prejudice.

Or consider gay gentrification. The emergence of a gay and lesbian movement since the late 1960s has translated geographically into the growth of numerous gay-identified neighborhoods (New York's West Village, San Francisco's Castro) and a few lesbian-identified neighborhoods (Castells 1983). In Rothenberg's (1995) story of gentrification in Brooklyn's Park Slope, the central theme is a community activism to provide the kind of neighborhood, services and access to housing that encourage the constitution of lesbian identities, singly and as a community. In a similar vein, Castells (1983) has shown in San Francisco that gentrification is a strategy for countering a housing market that broadly discriminates against gays and lesbians. But as Lauria and Knopp (1985) argue, the politics of gay gentrification are more complex, and theoretical analysis has to go further. The connection to gentrification is real enough but not all gay neighborhoods are gentrified. Larry Knopp (1989, 1990a, 1990b) has made the most sustained and insightful analyses of gay involvement in gentrification and finds generalizations difficult to make. Although gay and lesbian communities are multiracial and multiclass, they are still skewed significantly toward a white middle-class and comparatively well-off population. When this is combined with the social activism that helps construct gay and lesbian identity against social oppression, gentrification begins to seem like a geographical as well as a social strategy of identity construction. Gay gentrification in particular can therefore be a very conservative enterprise. Knopp insists that any understanding of gay involvement in gentrification must consider questions of the housing market along with gay identity construction, finding in New Orleans that there is significant gay entrepreneurship in gentrified real estate markets. Gay gentrification, he suggests (Knopp 1990a), is a strategy of combined class and gay identity construction.

The point of the argument, then, is that gentrification is an inherently class-rooted process, but it is also a lot more. At the regional, urban or community scales, the challenge is to understand the chain of connections linking specific local responses and initiatives to overall social structure, or indeed to determine when such links are too tenuous to be sustained (see also Katznelson 1981). The supposed rigidity of marxist conceptions of class is generally overplayed. Concepts of the working class and capitalist class are often interpreted with a perverse empiricist literalism; a single individual who does not quite fit every angle of the conceptual class mold is deemed sufficient excuse to consign to the dustbin the mold, its maker and all who have cast eyes on them. That certainly was not Marx's own conception of class and I seriously doubt that such a strict conceptual Calvinism characterizes even the more "functionalist" misinterpreters of Marx.

Nonetheless, to clarify any possible misunderstanding, it is worth restating a basic understanding of class. In the context of recent debates, it is Wright (1978) who points the way toward a better conceptualization of class. While accepting that class position depends first and foremost on one's relationship to the means of production – whether one owns companies or is owned by them – Wright emphasizes that this criterion by no means provides hermetically sealed class boundaries. On the contrary, many people occupy "contradictory class positions"; the source of contradiction is historically differentiated and might involve anything from the occupation of an individual, to the level of class struggle in a given period. Classes are always in the process of constitution, as indeed are genders and sexual identities (Bondi 1991a).

This is particularly important: during periods of diminished class struggle, class boundaries become more difficult to identify. In this way, it is worth noting, the question of consciousness is built into the definition of class, which is not at all to say that it *determines* class. Classes should be seen not as pigeon-holes, overdetermined sets with precise boundaries and exact binary rules for inclusion and exclusion. Rather, classes resemble fuzzy sets which are defined more or less sharply depending on social, economic, political and ideological conditions. This shifts the argument about class structure from fatuous debates about which individuals fit in which box to a more serious historical concern with the rise and decline of specific classes as such and their changing constitution.

But this loosening up of class categories is only part of the answer. It certainly opens up a space not only for the middle classes but also for a changing emphasis on different classes and class relations, depending on the scale and topic of analysis. There remains, however, the more far-reaching objection that a class analysis derived from the social relations of production is wholly inappropriate for comprehending the consumption sphere in which gentrification is seen to fit. In the context of urban change, this argument has been proposed most insistently by Peter Saunders. Saunders (1978, 1981) began with an attempt to rehabilitate the concept of housing class initially proposed by Rex and Moore (1967). Positioning himself, as Rex and Moore did, on squarely Weberian theoretical ground, Saunders argued that notions of class derived from marxist analysis were at best relevant only in the sphere of production, but did not apply to the "analytically distinct sphere" of consumption. Since consumption patterns, and especially patterns of housing consumption, are more important sources of social differentiation than is generally acknowledged, it was necessary, Saunders argued, to set up a parallel set of class distinctions based on the means of consumption. However, Saunders subsequently abandoned "the attempt to theorize home ownership as a determinant of class structuration" (1984: 202), as a result not so much of a critique of the Weberian origins of the notion as of a hardening of his separation between production and consumption spheres: his previous approach "overextends class theory and ultimately fails to relate class relations generated around ownership of domestic property to those generated around ownership of means of production" (1984: 206). Instead, he suggests that "divisions in the sphere of consumption do not restructure class relations but do crosscut them." Further, "consumption-based material interests are no less 'basic' or 'fundamental' than production-based (class) ones." Saunders (1990: 68–69) has gone so far as to suggest that this separate consumption sphere may be undergirded by a biological predisposition toward homeownership.

Now this critique of marxist class analysis and affiliation with Weber has had considerable appeal among researchers in gentrification, even if it leads to a separation of spheres which feminist analyses largely sought to integrate. In particular, it has provided a direct or indirect means by which explanations of gentrification can be rooted in contemporary social restructuring to the neglect of economic considerations. In a summary critique, Harloe (1984) has challenged what he sees as Saunders' asserted rather than demonstrated social relationships. Harloe especially challenges Saunders' contention that British society is witnessing "a major new fault" line drawn not on the basis of class (ownership of means of production) but on the basis of sectoral alignment (ownership of means of consumption) (Harloe 1984: 233). This new fault line, for Saunders, separates owners of residential means of consumption (homeowners) from those who do not own their means of consumption and are thereby forced to consume collective means of consumption; that is, state-owned or social housing. Privatization of housing does not constitute a new and long-term fault line, according to Harloe, notwithstanding the numerical decline of social housing in the 1980s. It should be added that Saunders' distinction makes even less empirical sense in the US where, unlike in Britain, approximately a third of all households still get their housing from the private rental market while less than 3 percent are public renters. These "nonowners of the means of consumption" do not represent a coherent consumption sector at all but are widely spread across the class map. Thus in New York City, 20 percent of renting households had an income of less than $4,960 in 1980 and 25.3 percent earned incomes below the official poverty level. But by the same token, 20 percent of renting households earned over $22,744 (Stegman 1982: 146–150). The top of the market is dominated by luxury rentals just as the bottom is dominated by public housing and rooming houses.

But the application of Saunders' critique presents special conceptual problems in the context of gentrification. While not everyone would openly subscribe to such a radical distinction between production and consumption as Saunders has proposed, the middle ground of a mutual interaction between consumption and production has not been explored in practice, and so researchers have tended to come down on one side of this dichotomy or the other. Thus I accept Hamnett's (1984) critique of my earlier work (N. Smith 1979a) for conflating a variety of lifestyle and demographic arguments under a somewhat grab-bag concept of consumption-side and consumer preference explanations. The attempt to integrate consumption-side and production-side arguments – not in some mechanical resort to the notion that one "crosscuts" the other but rather in the notion that production and consumption are mutually implicated – should be at the top of our agenda (see Boyle 1995 for a critique).

However, this might not prove as easy in practice as it is in words. Attempts to reformulate marxist analyses and to soften the exclusivity of the earlier emphasis on economic and production-side questions have emphasized a different set of questions (Rose 1984; Beauregard 1986; Williams 1986). These contributions have focused on the following questions: where do the "gentrifiers" come from? What are the social processes that are responsible for producing them as a coherent social group? As Beauregard (1986: 41) puts it: the "explanation for gentrification begins with the presence of gentrifiers" (see also Rose 1984: 55–57). The overarching importance of this work is that it has introduced the broad questions of social restructuring which are clearly fundamental to explaining gentrification. However, they also bring certain intrinsic dangers with them. If indeed gentrification is to be explained first and foremost as the result of the emergence of a new social group – whether defined in class, gender or other terms – then it becomes difficult to avoid at least a tacit subscription to some sort of consumer preference model, no matter how watered down. How else does this new social group bring about gentrification, except by demanding certain kinds and locations of housing in the market? This is implicit even in Saunders' approach, when he argues (after Giddens 1981) that homeownership fulfills some deep-seated desire for "ontological security" (1984: 222–223; cf. also Rose 1984).

I do not mean to suggest here that consumer demand is illusionary or that it finds no expression in the market. Nor do I mean to suggest that such demand is unchanging or impotent. There is no argument but that demand can at times – and especially those times when demand changes dramatically – alter the nature of production. But the conundrum of gentrification does not turn on explaining where middle-class demand comes from. Rather it turns on explaining the essentially *geographical* question why central and inner areas of the city, which for decades could not satisfy the demands of the middle class, now appear to do so handsomely. If indeed demand structures have changed, we need to explain why these changed demands have led to a *spatial* re-emphasis on the central and inner city.

We can now return to the hypothetical possibilities that emerged from the earlier discussion of the new middle class. Three possibilities were suggested which tied economic and labor-market transformations to gentrification. Whether or not the increase in incomes in the upper middle range amounts to the emergence of a new middle class, it is difficult to imagine how this economic argument could account for gentrification. Higher levels of disposable income may well make a central- or inner-city domicile affordable for large numbers of people, but this is at best an enabling condition. Higher incomes do not in themselves imply a spatial bias toward the central city; indeed quite the opposite assumption was an important cornerstone (erroneously, it now seems) of land use theory based in neoclassical economics (Alonso 1960). Equally, there can be little doubt that a continued and even accelerated centralization of administrative, executive, professional, managerial and some service activities may make a central domicile more desirable for a substantial sector of the middle class. But do these arguments really amount to an explanation of the geographical *reversal* of location habits by a proportion of middle-class women and men? Were there no young upwardly mobile professionals in the 1950s, or even the 1920s? Why did not the "proto-yuppies" of six or eight decades ago initiate the gentrification process instead of spear-heading the rush to the suburbs? Where younger middle-class people did reclaim city neighborhoods against the suburbanization trend, they were sufficiently uncommon that, as in Greenwich Village, they and their lifestyles were described as "Bohemian."

A similar argument can be made about the changing roles of women (Séguin 1989). Although more distinct economic or occupational changes are evident for at least some women, identified by class, and the data suggest a clear correlation with gentrification activity, the conspicuousness of these changes should not blind us to less visible but possibly more trenchant changes. The increase in the number of households with two or more incomes certainly enhances the rationale for a central domicile for those households with central workplaces. But it is incumbent on us to explain why the gradual quantitative increase in the proportion of women working and of women in higher income brackets translates into a substantial *spatial* shift of domicile. After all, married women were in the official workforce before (as well as during) World War II, albeit in smaller numbers, and some of these were in well-paying professional positions, yet no gentrification seems to have blunted the suburban flight of the time. How could such a comparatively quick spatial reversal be explained by more gradual social changes alone? And why, in any case, do such social changes lead to spatial changes in residence as opposed to an ossified spatial structure where increased costs are simply borne by the affected households? Of course Engels' "first law of dialectics," namely that quantitative change converts into qualitative change, might be invoked to account for such a shift, but I for one am skeptical about this "law". We need to give a more concrete explanation of the manner in which this quantitative social change suddenly becomes qualitative. Therefore as it stands, the argument that social restructuring is the primary impetus behind gentrification here is substantially "underdetermined."

It is not unreasonable to conclude, then, that Markusen (1981: 32) overstates the case when she suggests that "gentrification is in large part a result of the breakdown of the patriarchal household." The breakdown of the patriarchal household is undeniable, as is its contribution to gentrification, and this is a doubly important argument when linked to changing employment patterns. The breaking down of "sexual apartheid which kept men and women in rigidly separate spheres" (Samuel 1982: 124) seems to have selectively benefited women of higher socioeconomic position, white women, and more educated women, for whom some previously closed occupations have now become somewhat more available, the "glass ceiling" notwithstanding. And although a gradual loosening of this sexual apartheid may already have been occurring, it is predominantly the result of the strong feminist movement (Rose 1984) which emerged in the 1960s and influenced social legislation and social norms in succeeding decades. Here indeed we may see a more cataclysmic political and social change that contributes to the spatial shift associated with gentrification. Yet this is only a part of the story; however trenchant, it is an incomplete basis on which to rest responsibility for the massive economic, social and geographical restructuring that gentrification represents. We still require a broader explanation for the unprecedented geographical shift of investment associated with gentrification.

Social restructuring is a vital piece of the gentrification puzzle, but it makes sense only in the context of the emergence of a rent gap and a wider political and economic restructuring. In the early decades of this century the disinvestment from central urban areas throughout North America, Australia and even Europe was in its infancy, the suburbs only beginning to sprawl, and the link between social restructuring and the restructuring of the central urban environment virtually inconceivable. Not until major disinvestment had created the opportunity for profitable reinvestment and "slum clearance" and urban renewal had demonstrated its feasibility would gentrification begin. It is the existence of the rent gap in the center that facilitates the translation of more gradual social as well as economic processes into a spatial reversal of some residential, recreational and employment activities.

GENTRIFICATION KITSCH AND THE CITY AS CONSUMPTION LANDSCAPE?

Embedded in Saunders' (1984) argument is a discussion of historical phases of consumption patterns that we can begin to relate back to production and produced gentrification landscapes. He claims that in the last 150 years there has been a succession of three phases in the mode of consumption. According to this empirical generalization of the history – for he explicitly rejects the assumption that these phases are evolutionary – a "market" phase dominated the nineteenth century, but was superseded by a "socialized" mode of consumption, and ultimately in the late 1970s by a "privatized" mode of consumption. Thus for Saunders, the back-to-market privatization of most national economies in the late twentieth century is part of a major, long-term restructuring away from state involvement in the sphere of consumption. The geographical corollary of this argument is the claim that the "urban reform ideology" of the new middle class, "the present day counterparts of Veblen's leisure class," is fashioning a postindustrial city with a consumption landscape rather than a production landscape (Ley 1980; see also Mills 1988; Warde 1991; Caulfield 1994). The world of industrial capitalism is superseded by the ideology of consumption pluralism, and gentrification is a signifier of this historical transformation, inscribed in the modern landscape. An urban dream is coming to supersede the suburban dream of past decades.

There is something appealing about this conclusion. As we watch a magnificent Eton Center rise in Toronto, stroll along Melbourne's Lygon Street, explore the delicately gentrified medieval passageways of Schnoor in Bremen, watch an unprecedented gentrification strategy catapult Glasgow into a European "City of Culture," or witness a postmodernist Bonaventure Hotel dominate the architecture of central Los Angeles, our field sense tells us that the times are certainly changing and that indeed something like a bourgeois playground is being constructed in many downtowns. But does this merit the conclusion that urban form is now being structured by consumption ideologies and demand preferences rather than production requirements and geographical patterns of capital mobility?

It is hardly a new theme that we are on the verge of a "consumption society." The same notion has continually surfaced both in analytical and in utopian tracts, and was a staple of sociological discourse long before Bell's (1973) announcement of postindustrial society. In the postwar period alone, this theme appears in different guises – in Burnham's managerial revolution, Galbraith's affluent society and Whyte's organization man. David Riesman (1961: 6, 20), in *The Lonely Crowd*, gave it particularly sharp and optimistic expression when he placed the postwar US alongside the Renaissance, the Reformation and the Industrial Revolution, and declared that the postwar revolution involved nothing less than "a whole range of social developments associated with a shift from an age of production to an age of consumption."

Doonesbury

BY GARRY TRUDEAU

Plate 5.3 Gentrification-induced displacement (Doonesbury © 1985 G. B. Trudeau.

In order to pin down what is implied in this "age of consumption" and to improve our comprehension of the limits and potential of the consumption landscape which come in its trail, let us return to Riesman's age of production – Saunders' age of a market mode of consumption – and pick the story up there.

During the nineteenth century, the international expansion of capital was accomplished primarily through the appropriation of absolute surplus value and the consequent economic expansion in absolute geographical space (N. Smith 1984). This is the period dominated, according to Michel Aglietta (1979), by the "extensive regime of accumulation." The close of the nineteenth century, however, witnessed both severe crises of overaccumulation and an increasingly organized and militant working class prepared to fight for a whole array of demands, which, while having an economic basis, included not only workplace issues of wage rates, the length of the working day and overall conditions, but also housing questions, rent levels, and so forth. In response to these economic and social challenges to capital, whether from overaccumulation or wage laborers, the capitalist system underwent a broad sequence of transformation toward an intensive rather than extensive regime of accumulation. Absolute surplus value was superseded by relative surplus value and this meant a revolutionizing of the workplace, the rise of Taylorism and scientific management. But it also meant a revolutionizing of consumption relations and of the deep-seated social relations between capital and labor. The solution to problems of overaccumulation lay ultimately in the economic enfranchisement of the working class, who became a powerful magnet of consumption in their own right. It was this intensive regime of accumulation, which Aglietta refers to as Fordism (Gramsci 1971 edn.), that dominated the two decades of dramatic postwar economic growth, and it involved an unprecedented intervention by the state. The transition to an intensive regime of accumulation is also marked by the geographical transition from the absolute expansion of global capitalism to its internal expansion and differentiation, and the emergence of the classical pattern of uneven development (N. Smith 1984; Dunford and Perrons 1983).

This is a highly schematic summary of many complex societal changes, but the urban transformation wrought in the midst of this process is readily recognizable. At the urban scale it was a period of dramatic suburbanization in which the state actively sponsored working-class homeownership and decentralization (Checkoway 1980; Harvey 1977) and a whole array of expanded consumption patterns. Where national and colonial expansion had once held the key to solving problems of overaccumulation and disequilibria between production and consumption, the key was now found in an internal redifferentiation of geographical space. The "suburban solution" (Walker 1977, 1981) was part of this, but equally a part of the solution was the expansion of a massive welfare state and, in general, the development of a much more socialized economy, as described by Saunders. This involved a rapid expansion of collective consumption, in housing, health care, education, transportation and so on; civil society and the people who composed it were drawn to an unprecedented degree into the heart of capital – a real subsumption under capital in the consumption sphere – through mortgages, car loans, consumer debt, and the costs of college education. This is the process to which O'Connor (1984: 170–171) refers when he notes that individual needs are increasingly social but at the same time are increasingly fulfilled by means of commodity consumption; real social needs are in fact frustrated by the expansion of commodity "needs." But as a result of this historical process there was a dilution of cultural boundaries, a partial homogenization of consumption patterns across class boundaries, and in the US at least a comparatively quiescent working class, owing in part to the capital–labor accords reached in the late 1940s and early 1950s. The very real fuzzing of class boundaries that came to the US in this period was not replicated in Britain and elsewhere for another two decades but it came nonetheless; classes were not only contradictory, they were less distinct than in the past, and it was this reality which gave rise to so much of the optimistic expectation of a homogeneous middle-class society.

"Capitalism shifted gears," according to David Harvey (1985b: 202), "from a 'supply-side' to a 'demand-side' urbanization." This gave rise in the ashes of World War II to what he calls the "Keynesian city," which, with its demand-led urbanization, took on all the appearances of a "post-industrial city." Consumption was no longer the poor relation of production, in the economy, in people's individual lives, or in the production of geographical space, as it had been under the extensive regime of accumulation.

Most of the analysis of this economic and social restructuring has focused on the new roles of the working class because their integration into a mass consumption society is one of the historically important developments that distinguishes Fordism. But the ethic of consumption which accompanied this sea change was by no means restricted to the working class, or rather to certain sectors of that class – namely unionized and largely white workers. It was generalized throughout postwar society. The middle class shared in the spirit of the time as much as the better-off sectors of the working class, but this was a much less remarkable event since the middle class was already identified with comfortable consumption habits. This new structure of consumption norms was in turn made possible by considerable stand-ardization of commodity production, especially in housing and automobiles, although the former sector continued to lag behind the production of consumer durables in terms of technical and organizational advances in the labor process. According to Aglietta, this integration of large segments of the working class into a consumption ethic

> meant the creation of a *functional aesthetic* ("design"), which acquired fundamental social importance. . . . Not content to create a space of objects of daily life, as supports of a capitalist commodity universe, it provided an image of this space by advertising techniques. This image was presented as an objectification of consumption status which individuals could perceive outside themselves. The process of social recognition was externalized and fetishized.
>
> (Aglietta 1979: 160–161)

Just as the standardization and cheapening of commodities extended the consumption habits of the working class, it made more and different commodities available to (and even prized by) the middle class; the standardization of products at one pole placed a particular premium on differentiation at the other. It is this question of cultural differentiation in a mass market which is most relevant to gentrification. Gentrification is a redifferentiation of the cultural, social and economic landscape, and to that extent one can see in the very patterns of consumption clear attempts at social differentiation. Thus according to Jager (1986; see also Williams 1986), gentrification and the mode of consumption it engenders are an integral part of class constitution; for Jager they are part of the means employed by new middle-class individuals in order to distinguish themselves from, on the one hand, the old middle class, and, on the other, the working class. Interpreting "urban conservation" in several Victorian neighborhoods in Melbourne, he writes:

> urban conservation is the production of social differentiation; it is one mechanism through which social differences are turned into social distinction. Slums become Victoriana, and housing becomes a cultural investment with facadal display signifying social ascension. . . . The ambiguity and compromise of the new middle classes is revealed in their aesthetic tastes. It is through facadal restoration work that urban conservation expresses its approximation to a former consumption model in which prestige is based upon a "constraint of superfluousness". . . . But in the case of urban conservation those consumption practices are anxiously doubled up on what may be termed a Victorian work ethic embedded in renovation work. In artistic terms this duality is expressed as that of form and function. . . . The effacing of an industrial past and a working class presence, the white-washing of a former social stain, was achieved through extensive remodelling. The return to historical purity and authenticity (of the "high" Victorian era) is realised by stripping away external additions, by sandblasting, by internal gutting. The restoration of an *anterior* history was virtually the only manner in which the recent stigma of the inner areas could be removed or redefined. . . . Inner worldly asceticism becomes public display; bare brick walls and exposed timbers come to signify cultural discernment, not the poverty of slums without plaster. . . . In this way "the stigma of labour" . . . is both removed and made other. Remnants of a past English colonial presence survive through the importance attributed to hand made bricks, preferably with convict thumbprints.
>
> (Jager 1986: 79–80, 83, 85)

The pursuit of difference, diversity and distinction forms the basis of the new urban ideology but it is not without contradiction. It embodies a search for diversity as long as it is highly ordered, and a glorification of the past as long as it is safely brought into the present. If it is in part a reaction to the perceived homogeneity of the suburban dream (Allen 1984: 34), the urban dream also shares much of the angst of the former. History was a commodity no less for Connell's (1976) commuting restorers of Surrey farm cottages than for Jager's

Melbourne would-be Victorians. Whether in the urbs, burbs or exurbs, however, the perpetual search for difference and distinction amid mass consumption is eternally frustrated. It can lead to a new "gentrification kitsch" in which cultural difference itself becomes mass produced. As the process proceeds, this becomes increasingly clear. As the choicest structures are converted and open sites become increasingly conspicuous, as well as expensive, in otherwise gentrified neighborhoods, the infill is accomplished by new construction. Here the architectural form provides no historical meaning that can be reworked into cultural display, and the appeal to the kitsch of gentrification is therefore more extreme. Where such modern infill occurs in gentrifying neighborhoods, whether in Baltimore's Otterbein, in central London Barrett estates, or in Brisbane's Spring Hill, the impression is one of having come full circle, in geographical and cultural as well as architectural terms. This infill gentrification is accomplishing a suburbanization of the city.

How does this perspective differ from that of postindustrial advocates who now see a demand-led urbanization in which consumption landscapes are superseding production landscapes? The answer comes in three parts. First, the perspective implied above attempts to understand the changing importance of consumption in terms of the social restructuring that has occurred since the end of the nineteenth century, and, in turn, to relate that social restructuring to the economic and spatial restructuring which accompanies it.

Second, urban development in the postwar period may have been consumption led in the sense that the accumulation of capital depended on an unprecedented level of production of consumption goods, and that this aspect of the economy had a highly spatial identity. It is necessary to be careful about this argument, however, because postwar expansion also witnessed massive increases in the production of means of production, not to mention means of destruction. The latter was also a defining feature of postwar growth, as suggested in the notion of a permanent arms economy (Vance 1951). In any case, "consumption-led urbanization" is not necessarily the same thing as "demand-led urbanization," and here I think Harvey (1985b) conflates the two in his discussion of the Keynesian city. Consumption-led growth implies the importance of the consumption sector of the economy and the production of goods in that sector, whereas "demand-side urbanization" implies that in the move from the extensive to the intensive regime of accumulation, the dynamics and demands of accumulation are now subordinated to those of consumption. Accumulation is potentially relegated to a by-product of consumption. Some sort of consumer demand theory would then be necessary to understand the direction of "demand-led urbanization."

The third, and arguably most important, difference between the perspective proposed here and the postindustrial thesis is essentially historical, and suggests a justification for the accumulation model elaborated above. If postwar expansion, the political pact between labor and capital, and a developing ethic of mass consumption substantially muddied the waters of class distinction during this period – in reality as well as in appearance – this was not destined to be a permanent, gradual and progressive transformation of advanced capitalist societies into a homogeneous middle class. The year 1973 was an ironic one for Daniel Bell, the 1950s author of *The End of Ideology*, to publish his tract on postindustrial society, because it was in the same year that the industrial system began to reassert itself with a vengeance. For several years, the international economic system had been stumbling toward recession, but the oil embargo of October 1973 triggered a crisis – reaching much deeper into the social fabric than a mere energy crisis. This crisis in turn initiated a broad-based economic, political and social restructuring that had become global by 1989 and is still being worked through two decades later. The mutual reconfiguration of consumption and production after 1973 had dramatic spatial effects.

To the extent that urbanization and changing urban form were part of the solution to the problems of an earlier regime of capital, they just as quickly became part of the problem when the intensive regime broke down. The sharp changes in urban patterns since the late 1960s suggest the extent to which the city will necessarily be part of the solution to contemporary crises (Harvey 1985b: 212). The optimistic homogenization of society in the realm of consumption was only partially accomplished, and after the 1980s has been abruptly truncated and even reversed; the social equalization championed by consumption-led urbanization has given way to a hard social redifferentiation along class and race lines at the hands of continued industrial decline, high chronic rates of unemployment, and the wide-scale privatization of any public infrastructure that could be sold. The dismantling of the welfare state systems erected in the first decades of the century represent a reversal not only of the gains achieved by the feminist and civil rights movements in the 1960s, but of the gains made by the postwar left and Keynesian liberals before them.

The point here, as regards the social dimensions of gentrification, is that the economic restructuring of the 1970s onwards has been accompanied by a social restructuring in which a new cleavage is being asserted. The "new fault lines" in part reassert old class lines, but, with the expansion of the so-called service sector, they also cut into new territory. The overall result is an increasingly polarized city; about this there has been consensus from right as well as left (Sternlieb and Hughes 1983; Marcuse 1986). That is, the consumption ethic and consumption-led urbanization have continued to be a reality for many in the middle classes while they are a "dream turned sour" for most industrial and service workers.

Throughout the recession of the late 1970s and early 1980s, in most industrialized economies gentrification proceeded apace (on Canada see Ley, 1992) while homelessness escalated. Not until the depression of the late 1980s did the urban dream begin to tarnish. This suggests a radical bifurcation of the consumption dream, producing "a city of haves and have-nots" (Goodwin 1984). The leading historical edge of the "city of the haves" is well represented by the so-called yuppies and burgeoning gentrification; round the corner and in close geographical proximity is the city of the "have-nots," represented by as many as 3 million people in the US alone. The dismantling of public services and privatization of public functions since the mid-1980s has given gentrification an altogether more ominous social meaning.

Part II

THE GLOBAL IS THE LOCAL

MARKET, STATE AND IDEOLOGY

Society Hill

Emergent gentrification in the late 1950s and early 1960s quickly earned a symbolic currency that surely overreached its economic and geographical significance "on the ground." A small but highly visible outlet for productive capital seeking a profitable resting place, gentrification seemed to promise a reversal of postwar residential decline and decentralization. Ideologies of gentrification quickly fastened on healthy neighborhoods where once there had been decay, profit where there had been poverty, the middle class back in the city: gentrification was "a good thing." First conceived in the late 1950s, the gentrification of Society Hill in Philadelphia was an especially "good thing" and especially symbolic. Set between the Delaware River and Center City, the neighborhood occupied the site of William Penn's seventeenth-century "holy experiment": lying immediately to the south of Independence Hall, the Mall and the Liberty Bell, Society Hill was widely touted in Philadelphia tourist and historic preservation literature as part of "the most historic square mile in the nation" (Figure 6.1). By the late 1960s, this neighborhood of late-eighteenth- and early-nineteenth-century town houses had been repackaged both in the media and in academic urban geography, urban studies and sociology literatures as a centerpiece of a Philadelphia "renaissance" which, "since the 1960s," according to one writer, "has been the most widely illustrated example of on-going comprehensive restructuring and systematic renewal of an historic urban core" (Morris 1975: 148).

The gentrification of Society Hill was brought about by an intricate intertwining of state and financial institutions together with an early and influential prototype of the public–private development corporation. The context was the 1950s. Postwar economic expansion funneled capital toward the development of the suburbs and only very selectively toward existing urban centres. Whatever its distinguished past, Philadelphia more and more resembled a decaying east-coast industrial city in the minds of the ruling-class white elite. Major postwar urban renewal legislation was recently initiated at the federal level and the emerging model called for widespread slum clearance and urban renewal. The commitment to the rehabilitation of Society Hill's historic but crumbled housing stock was, in its time, a significant departure from renewal practices that would soon be pilloried in the popular press as much as the academic urban literature (cf. Anderson 1964).

Society Hill is always celebrated as the original home of Philadelphia's gentry, beginning in the seventeenth century and lasting to the Civil War. And

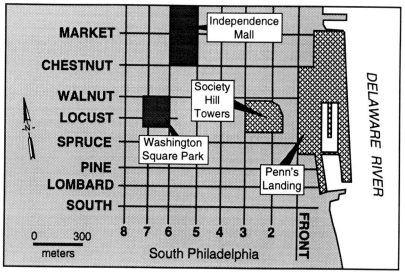

Figure 6.1 Society Hill, Philadelphia

since this was a slave-owning class, it was also home to an African-American community for just as long. With industrialization clustering around the adjacent Delaware waterfront (including the establishment of the city's major food market) and in neighboring South Philadelphia, and with the resultant urban growth in the second half of the nineteenth century, the city's upper classes moved west to Rittenhouse Square and to the first suburban communities over the Schuylkill River. African-Americans remained. This migration initiated a sequential disinvestment in Society Hill's housing stock that lasted for the best part of a century. By the 1950s many buildings lay vacant, abandoned by landlords, while others provided miserable, cramped, substandard accommodations for poor white and black working-class residents. During that decade, Society Hill's population fell by more than a half to 3,378, 21 percent of whom were "nonwhite"; the area lost 18 percent of its housing units, and a further 13.2 percent were vacant. An urban renewal plan was first drafted in the late 1950s, and from the start it involved public, quasi-public and private institutions. Its twin objectives were a revival of the city economy and the attraction of rich households "back from the suburbs."[1]

Plate 6.1   Society Hill, Philadelphia, on the eve of gentrification
(Urban Archives, Temple University)

THE RECIPE AND THE RHETORIC

From the start, Society Hill was seen as a "new urban frontier," and the gentrifiers as proud "pioneers" (Roberts 1979; Stecklow 1978). Initiated in 1959, the gentrification of Society Hill achieved instant celebrity status. According to an analysis by the influential Albert M. Greenfield and Co., "many planning authorities view[ed] Society Hill as one of the foremost renewal undertakings in the United States" (Greenfield and Co. 1964: 10). Albert Greenfield was not only "the largest and richest real estate operator in the city" (Burt 1963: 10) but a former chairman of the City Planning Commission, and was presumably in a position to know. If, given his professional and financial interest in the success of Society Hill, we might reasonably be wary of exaggeration, we have no such cause to be wary of the novelist Nathaniel Burt (1963: 556–557):

> The plan, now actually being put into effect, is one of the most daring and most tasteful pieces of town planning ever conceived, an attempt to salvage what is good of the old, add what is needed of the new, and in general transform that part of the city into a sort of urban residential paradise without making a museum-fossil out of it. When and if it all does get done according to plan, Society Hill will be an American showplace of city restoration.

Indeed, by the 1970s Society Hill had become just such a showplace with its tree-lined streets of red-brick Colonial and Federalist buildings, herringbone pavements, wrought iron streetlamps, and heavy stained front doors. So successful was the rehabilitation project, in fact, that residents quickly began to organize against the opening of any further commercial establishments which on weekends would bring yet further throngs of tourists into the neighborhood.

Society Hill's recipe for gentrification contained three essential ingredients: a private–public organization called the Greater Philadelphia Movement (GPM) and its offspring, the Old Philadelphia Development Corporation (OPDC); the state, at federal and city levels, mainly; and private financial institutions. Formed in 1952, GPM was no mere pressure group. According to J. S. Clark, who was elected mayor of Philadelphia in the same year, GPM was "predominantly a group of conservative but intelligent businessmen of integrity who have the interest of their city very much at heart" (quoted in Adde 1969: 35). Its membership list combined a who's who of Old Philadelphia families with newer corporate and public bureaucratic aspirants to the ruling class. Their aim was nothing less than the physical and financial revitalization of the entire city. The initial catalyst to gentrification, GPM immediately focused on Society Hill as the vital core of a "Philadelphia renaissance," and it perceived the city's food market on Dock Street (now the site of Society Hill Towers), along the eastern edge of Society Hill, as a major obstacle. It used its political clout to ensure the removal of the food market out to South Philadelphia in 1958, and thereby made the fledgling renewal plan a serious proposition.

In the meantime, the Greater Philadelphia Movement, which preferred to keep a genteel distance from the nitty-gritty of political implementation, spawned an organization which it felt was more suited to the overseeing of redevelopment. The task of the OPDC was more practical and hard-headed. It liaised between local government, planners, investors, developers and home-owners, and promoted Society Hill in the local and national media.[2] If OPDC gave a populist spin to its vision, emphasizing that the city's revitalization was a "community project" involving everyone, its focus firmly connected a gentrified Philadelphia with the profitable futures of their firms. In the words of William Kelly, a former president of the First Pennsylvania Bank:

> The future of our companies, all of them, is tied to the growth of our city. When I spend time on civic affairs I'm in effect working on the bank's business too. . . . The growth of our bank, its well-being in the years to come, depends on what is done here in Philadelphia.
>
> (quoted in Adde 1969: 36)

Kelly might have added the converse: that revitalization of the city and its well-being depends on the cooperation of the banks, and it in turn stimulates *their* growth.

If GPM and OPDC provided the lubricant between the will of the state and the resources of the private sector, the private sector was nevertheless vital. Not only had private disinvestment set up the opportunity for reinvestment in the first place, but reinvestment by financial institutions would supply the bulk of reinvested capital in the form of mortgages and loans. These went to individual restorers as well as larger professional and corporate developers. From virtual redlining in the early 1950s, the neighborhood went in fifteen years to what would later be known as "greenlining": as we shall see, financial institutions increasingly sought to invest wherever they could in Society Hill.

The state, broadly conceived, constituted the third major ingredient in Society Hill's gentrification, acting variously as an economic, political and ideological agent for the project. Inspired by a poetic vision of Philadelphia's revival, the City Planning Commission devised the renewal plan. More than to anyone else, the vision belonged to the Commission's executive director, Edmund Bacon, who had himself been a member of GPM. More practical than visionary, the city government accepted the Commission's plans, made the necessary rezoning changes, and provided 30 percent of the state's project costs. As was the traditional practice at the time, the city government created an entirely new "local public agency" to implement this and other urban renewal schemes. The "Redevelopment Authority of Philadelphia" was run by a board whose members were appointed by the mayor; its funds came predominantly from federal sources as well as the city, and to a lesser extent the state of Pennsylvania.

The federal government's role was twofold. The gentrification of Society Hill was actually organized as a result of (and under the provisions of) the Housing Acts of 1949 and 1954. The 1949 Act established the basic legislation for fed-eral involvement in urban renewal and what became widely known as the "Title 1" provisions. The 1954 Act, among other things, provided for the reha-bilitation of buildings (not just "slum clearance") as part of urban renewal,

Plate 6.2 The Dock Street market, Philadelphia, in the 1940s
(Urban Archives, Temple University)

and, as an amendment to Title 1, this was crucial for Society Hill. Under the legislation, the federal government provided 67 percent of the project costs, but since a condition of this legislation was that no Title 1 project should cost over $20 million in public funds, the Society Hill project was renamed and split into three separate units, christened respectively "Washington Square East Urban Renewal Area, Units 1, 2 and 3." The second aspect of the federal government's involvement came with FHA-insured mortgages provided to a number of developers in Society Hill, particularly under Section 312, which financed urban "homesteading."

Although for analytic purposes it is convenient to view the public–private organizations, financial institutions and state bodies as distinct agents, in reality they were not. William Day, for example, president of OPDC in the late 1960s, was also board chairman of the First Pennsylvania Banking and Trust Co., which invested large amounts in Society Hill. He was succeeded to the presidency of OPDC by William Rafsky, director of the Redevelopment Authority, who also came to head Philadelphia 76 (the City-appointed group which organized Philadelphia's Bicentennial celebrations). But perhaps the most notorious case of overlap was Gustave Amsterdam. In the late 1960s Amsterdam was executive director of the Redevelopment Authority and executive vice president of OPDC. He was also Chairman of the Bank Securities

Corporation, a private financial enterprise which financed one or more Redevelopment Authority contracts held by a building firm with which he was also connected. When it was discovered in 1969 that he had used his public and quasi-public positions to enhance his private investments, he was forced to resign. This happy confusion of ruling-class interest and public largesse was no mere imperfection of an otherwise perfect plan; it was written into the rhetoric from the start and seen as fundamental to the very concept of Society Hill.[3]

STATE CONTROL

The state had both a political stake in realizing Society Hill and an economic role in helping to produce this new urban space. In implementing the plan, the Redevelopment Authority's primary responsibility was political control. In 1959 it began acquiring all the properties in Unit 1 of the renewal area. Invoking widespread authority of eminent domain and newly created building "conformity" codes, the Redevelopment Authority gave tenants two months to leave. Cursory offers of relocation assistance were only sporadically adopted by tenants. To property owners, it offered "fair market price," which for some represented an unprecedented windfall insofar as their buildings were virtually unsellable, while for others it amounted to an unceremonious taking of property. The Redevelopment Authority proceeded where necessary to demolish buildings that were structurally unsalvageable, otherwise improved the sites, then resold properties and sites to designated developers at the "appraised site value." The Authority, in other words, absorbed the costs of converting lived and abandoned housing into redevelopment sites. The Authority also exercised control after the sale of a building or site was complete. To acquire a property, all developers were required to enter a legal agreement with the Authority, stipulating a building's structure, external architecture and function, and the date by which redevelopment would be completed. Defaulters were liable for prosecution, and numerous cases were indeed pursued in the courts by the Redevelopment Authority.

Part of its political control the Redevelopment Authority eventually delegated to OPDC. With a rising demand for historic houses, subsequent to the successful hyping of Society Hill, the Authority found itself spending increasing amounts of time just selecting individual developers. In 1967 it handed this task over to OPDC, giving it an initial portfolio of 190 properties. According to OPDC president and Redevelopment Authority director William Rafsky, the Corporation nominated developers according to three criteria: they had to demonstrate the financial ability to rehabilitate, the average cost of which in the early 1970s was around $40,000 (Old Philadelphia Development Corporation 1975); architecture sympathetic to the historic character of Society Hill was "preferred"; and plans for single-family owner-occupied dwellings were also preferred. OPDC did not advertise properties. Interested developers, it was assumed, would hear through word of mouth, or private connections, or else would simply be inspired by glowing media accounts of the civic beneficence of Society Hill restoration.

The Redevelopment Authority's political control over the project was closely interlocked with its economic role, but here a somewhat converse aspect of the relationship between capital and the state begins to emerge. By July 31, 1976, by which time the project was effectively closed out, $38.6 million of federal and city funds had been invested to ensure the success of Society Hill, virtually all of it funneled through the Redevelopment Authority. Much of the initial money was raised by municipal bond issues purchased by the city's largest banks and financial corporations. In other words, the largest financial institutions financed the state at zero risk to invest in an area where these same institutions would not get involved directly themselves without the state's involvement. They would then consider doing business. Put this way, of course, it is difficult to avoid the conclusion that state involvement was little more than a catalyst for the financial interests of the city's corporate elite and that gentrification was as much a workable means toward urban profitability as toward livability. From this point of view, the purpose of the state was to re-create the profitability of urban real estate. Where the private market had profited by the disinvestment from Society Hill, the state was now being required to invest funds to amortize the disinvestment so that the same neighborhood could be made profitable again for private reinvestment.

Of the $38.6 million of public funds invested by 1976, approximately $4.2 million was spent on surveying, legal services, interest payments, administrative costs, and the like. The net cost of property acquired and resold by the Authority was $23.6 million, representing a state expenditure of "property capital" (Lamarche 1976); the remaining $10 million was productively invested on demolition, clearance and site improvements.[4] These expenditures represented direct subsidies to the developers, who not only faced proportionately lower redevelopment costs but also absorbed the surpluses and profits produced by workers in the public sphere.

SOCIETY HILL'S DEVELOPERS

Three kinds of developer were attracted to Society Hill. All were constrained on the one hand by the design established by the state and public–private institutions and on the other by the need for mortgage financing and construction loans from the private sector. They were:

(a) Professional developers who bought property, redeveloped it, and resold it for profit;
(b) Occupier developers who bought and redeveloped property but lived in it after completion; and
(c) Landlord developers who bought and redeveloped property, and rented it to tenants after completion.

Landlord developers ranged from the single-property landlord and the professional landlord with numerous properties, to Alcoa (Aluminum Corporation of America), which, when its Society Hill project was completed in 1964, had $300 million invested in property (Kay 1966: 280). Alcoa in fact was simultaneously landlord and professional developer. A Pittsburgh-based multinational corporation and the world's largest aluminum concern, Alcoa was looking to diversify its operations as a hedge against the declining profit rates that began to affect extraction-based metal industries in the early 1960s. Real estate investment would provide high depreciation allowances in its tax returns as well as turn a high profit rate, so Alcoa became involved in the plan to build three thirty-storey towers with a total of 703 luxury apart-ments (together with thirty-seven low-rise town houses). Architecturally, the towers dominated the Delaware waterfront south of downtown. They were designed by the renowned architect I. M. Pei, who described the apart-ments thus: "The dwelling units themselves will be modern, air conditioned, and of ample dimensions, with rooms exceeding FHA [Federal Housing Administration] standards by comfortable margins."[5] In early 1978 the monthly rents ranged as high as $1,050 for a four-bedroom suite – a rent level that easily placed Society Hill Towers toward the top end of Philadelphia's luxury housing market.

Alcoa's involvement in Society Hill, as reconstructed mainly from Redevelopment Authority records, is a fascinating tale of real estate dealing. Alcoa became involved through its partnership with Webb and Knapp – a New York-based property company, owned and controlled by William Zeckendorf Sr., with assets approaching half a billion dollars. This probably made Zeckendorf's property empire the country's largest at the time. Certainly it was the most renowned. At different times he owned Manhattan's Chrysler Building and Chicago's Hancock Building; he built the Denver Hilton and Washington, DC's L'Enfant Plaza, and he assembled the land for New York's Chase Manhattan Plaza and, in earlier days, the UN Building (Downie 1974: 69–74).

Plate 6.3 Ground clearance for Society Hill Towers, 1961
(courtesy of Urban Archives, Temple University)

In May 1961, Webb and Knapp purchased the land designated for Society Hill Towers from the Redevelopment Authority. The company paid $1.3 million and to finance construction it proceeded to secure a 3 percent FHA-insured mortgage for $14.5 million under section 220 of the 1954 Housing Act. Alcoa, already a junior partner in the enterprise, bought out Webb and Knapp's interest in November 1962 when the developer suffered a periodic but severe short-term cash flow crisis. Through numerous incorporated subsidiaries inherited from Zeckendorf, Alcoa now held 90 percent of the contract for the Towers. The other 10 percent was owned by a British property development concern —Covent America Corporation – which was also a Zeckendorf partner. Such international partnerships in urban development were only beginning to come into existence as European capital, strengthened by postwar reconstruction, began to seek American investment opportunities in earnest. In any case, Society Hill Towers was completed and received its first tenants in 1964. Alcoa did not retain the building, however; it was resold in 1969 when the seven-year "double balance depreciation" period had expired and it no longer functioned as a tax write-off for the aluminum company. The new buyer was General Properties, a property company again associated with Zeckendorf. When, in the mid-1970s, General Properties' seven years of accelerated depreciation had in turn passed, they were confronted by a depressed property market with few prospective buyers. The owners attempted, instead of selling the Towers to a single buyer, to offload them onto the tenants as a tenant-owned cooperative. Too few tenants were willing to buy, however, and in 1976 General Properties finally managed to find a buyer in US Life, a Texas-based insurance company. And so the story continued. Through all of these tax-induced changes in ownership, the buildings were managed by Albert M. Greenfield and Co.

As the evolution of the Society Hill Towers illustrates, the gentrification of Society Hill was intimately tied into the rhythms and contours of wider national and international circuits of capital. Ownership and development interests were variously located from New York to Texas, Pittsburgh to London; the prime activities of the owners ranged from aluminum manufacturing to property development to life insurance; and the rationale for ownership had more to do with below-the-line profits and tax-reduction strategies than with a remake of the "most historic square mile" in the country.

And yet the gentrification of Society Hill was widely sold as a project of local "revitalization." If the Towers symbolize new construction by multinational capital, and the involvement of professional and landlord developers, the popular ideology of Society Hill emphasized the "occupier developers" – individual gentrifiers. The case of C. Jared Ingersoll and his wife Agnes was given early and prominent attention in the media as a means of boosting the area and igniting gentrification. Boasting an ancestor who signed the Declaration of Independence, Ingersoll was the scion of one of the city's "top families." As E. Digby Baltzell (1958: 311) put it in his study *Philadelphia Gentlemen*, even prior to the Society Hill affair, "the Ingersoll family . . . usually initiates the fashionable thing to do in Philadelphia." And so it was that the Ingersolls were persuaded to renovate a Society Hill "town house."

Plate 6.4 Society Hill Towers and townhouses, 1971
(Urban Archives, Temple University)

Restoration was presented as something of a civic duty for Old Philadelphia
patrician families, whom the Ingersolls represented *par excellence*. C. Jared had
been a director of US Steel and a member of the Philadelphia City Planning
Commission when in 1959 he and his wife agreed with OPDC and the
Redevelopment Authority to a highly publicized "restoration" in the newly
declared "Unit 1" of the Washington Square East urban renewal area. They
purchased a building shell on the once grand Spruce Street for $8,000 and
began a total rehabilitation of its Federalist facade and roomy interior.
With work completed at a cost of $55,000 and the building restored to
pristine eighteenth-century condition, they moved into their rehabilitated
town house at 217 Spruce in January 1961.

The Ingersolls were widely credited as the symbolic instigators of the
Society Hill "revival," an impression enhanced by Agnes Ingersoll in an article
for the Bryn Mawr alumnae magazine. After this symbolic move, a "Society
Hill restoration" (Ingersoll 1963) did indeed become all the fashion: it was
presented not simply as a civic duty but as something of a parlor game for Old
Philadelphians. At this point, Society Hill gentrification shared more with
Susan Mary Alsop's Georgetown of two decades earlier (Dowd 1993) than
with the process it helped initiate. But with public financing as well as private
financial backing, Society Hill flourished and was too lucrative for too many

people to let it be restricted to Philadelphia blue-bloods. Work officially began on Washington Square East Units 2 and 3.

FINANCING SOCIETY HILL

In addition to the state's role and that of developers, OPDC spearheaded an effort to convince banks and other financial institutions in the city to reverse their traditional redlining posture in Society Hill. These institutions played a vital role in providing mortgage and construction financing for the house-by-house reinvestment in the area, and this is revealed in greater detail in an analysis of mortgage lending activities (Tables 6.1 and 6.2). The figures refer to occupier and small landlord developers; that is, excluding large landlord and professional developers such as Alcoa.

Four discernible periods can be identified in the history of disinvestment and reinvestment at the heart of the gentrification of Society Hill. These are not rigidly defined, of course, but rather represent overlapping stages in the evolution of a gentrified landscape.

Pre-1952   Investment during the immediate postwar period was small scale and erratic. As real estate developers Albert M. Greenfield and Co. (1964: 16-17) put it:

> Financing in the area showed all the characteristics of a high risk neighborhood: secondary financing was common; the typical conventional mortgage showed a 50 to 60 per cent ratio; and there was a large number of private lenders, finance companies and mortgage lenders, specializing in high risks. . . . Investors and speculators were subdividing houses into small, substandard apartment units.

"Investment," referring to this period, is a euphemism. The predominance of high-risk investors, low mortgage ratios (the ratio of mortgage to purchase price), speculation and subdivision provide a classic portrait of concerted *dis*investment in the area. Little if any capital was invested productively in Society Hill. Larger, more stable lending institutions (including the state) were busy with low-risk, high-profit mortgages in the suburbs and even loans abroad. By refusing to lend capital for productive investment, not to mention the simple buying and selling of homes, the financial institutions contributed to the maintenance of the area as a "slum" as existing capital materialized in the built environment was further devalued.

1952-1959   With the well-publicized formation of GPM in 1952 and the focus of its attention toward Society Hill, interest in the area's investment potential was aroused. In 1954, as shown in Table 6.1, a third of all mortgages came from private sources. Private mortgaging can be a sign of affluent house buyers unconstrained by the need to take out a mortgage, but since the return to Society Hill was not yet under way, and since speculation was known to be rife, this latter cause is more likely. That an additional seventeen properties were purchased without mortgages (Greenfield and Co. 1964: 45) is a further

Table 6.1 The origin of mortgages in Society Hill, Philadelphia, for 1954, 1959 and 1962

Year	S&L	Banks	Insurance co.	Other insts.	Private	Unidentified	Total
1954	16	5	4	9	17	4	55
1959	29	6	3	0	15	2	55
1962	36	8	0	2	0	2	48

Source: Albert M. Greenfield and Co. 1964: Ch. 3. Figures for 1962 are based on half-yearly figures for January to June

Note: S&L = savings and loan institutions

Table 6.2 The origin of mortgages in Society Hill, Philadelphia, from 1963 at three-yearly intervals

Year	HUD/FHA	Fed. S&L	S&L	Comm. savings bank	Insurance co.	Finance corpns.	Savings fund	Community assn.	Private	Unidentified	Total
1963	0	12	6	8	0	0	0	1	1	51	79
1966	1	12	5	15	2	1	3	1	5	31	76
1969	4	11	1	7	3	2	4	0	3	21	56
1972	2	12	2	22	0	5	1	1	0	12	57
1975	1	9	5	16	3	6	3	4	4	10	61

Source: Redevelopment Authority of Philadelphia, Washington Square East Urban Renewal Area, Units 1, 2 and 3. Files: Philadelphia Real Estate Directory

Notes: HUD = Department of Housing and Urban Development; FHA = Federal Housing Administration; S&L = savings and loans institutions

*Plate 6.*5 Eighteenth-century housing stock, Philadelphia, before gentrification
(Urban Archives, Temple University Press)

Plate 6.6 Restored buildings at the corner of Second Street and Pine Street,
Philadelphia, 1965 (Urban Archives, Temple University)

indication of strong but small-scale speculation fueled by optimistic talk
in high places. Speculative buying and selling remains pronounced into 1959,
but the increased involvement of savings and loans institutions at the
expense of smaller, higher-risk lenders suggests a relaxation by medium-sized
institutions in their policies toward Society Hill. Although GPM's civic-minded
agitation stimulated speculation, then, it did not yet convince the larger banks
to enter the market, nor did it result in substantial investments of productive
capital. Government action, not their own words, was what bankers wanted
to see. Not until the state's renewal plan was implemented at the end of 1959
was capital reinvested in any quantity.

1960-1965 Once the redevelopment plan was implemented, and the
symbolism of the Ingersoll gesture had resonated among the ruling classes, it
did not take long for the largest institutions (banks and federal savings and
loans institutions in particular) to achieve a virtual monopoly in mortgage
lending. There was a simultaneous decline in mortgages from high-risk
institutions, which were forced elsewhere. Speculative capital too was substantially reduced – in large part, no doubt, because of the Redevelopment
Authority's strict control over rehabilitation once it took control in 1959. By
1963 the money flowing into Society Hill in the form of mortgages was

predominantly productive capital intended for building rehabilitation. Thus the mortgage ratio was over 200 percent in Unit 1, suggesting the dramatic extent of renovations; the average cost of a property was $13,124 while the average mortgage taken out on these properties was $26,700. Normal bank mortgages for housing purchase were between 80 percent and 90 percent of sale price. Thus the banks gave a virtual open line of credit to Society Hill's gentrifiers. By 1965, work had begun in all three units of the renewal area.

1966-1976 During this final stage, commercial and savings banks became the largest single source of mortgages. Their predominance is even more marked than the figures suggest, owing to the increased role played by real estate investment trusts, many of which were "affiliated" with the large banks. They are classified in Table 6.2 under 'finance corporations.' Society Hill had become a prime investment opportunity and its development was now led by the largest, most established and generally most conservative financial institutions. Theirs was the crucial role in sustaining development in this period, just as it had been in sustaining underdevelopment before 1959. Mortgages in this fourth period often exceeded $50,000; the risk was low and competition was fierce, which may explain the reappearance of various smaller institutions in the interstices of the market. Their appearance, however, did not seriously threaten the monopoly of larger institutions, and can perhaps be explained as a result more of preexisting personal connections with these smaller institutions rather than policy. The unusually low interest rates for these loans supports this interpretation.

By the mid-1960s, therefore, the large financial institutions had displaced the state as the primary economic and political dynamic behind the gentrification of Society Hill. This was as it was meant to be. In 1966 the mortgage ratio had declined to 142 percent, suggesting that while much productive capital still flowed to Society Hill for purposes of renovation, the process had already peaked. Units 2 and 3 increasingly attracted the productive capital as Unit 1 neared completion. Unit 1's mortgage ratio declined to 116 percent in 1972 and 54 percent in 1975, suggesting that buyers by this period were sufficiently affluent to make significant downpayments for the now largely rehabilitated and expensive stock. The average mortgage in Unit 1 by 1975 was $46,573, the average house price $86,892. With the plan essentially complete, the planners basking in the warmth of professional prestige, and the residents lounging in Federalist opulence, mortgages no longer financed production; they financed individual consumption.

CLASS, CONTEXT AND HISTORY

The Society Hill story is, as its boosters unfailingly stress, unique in many ways, but it also shares much with other experiences of gentrification. In social terms, its earliest conception was virtually aristocratic in origin, emanating largely from the drawing rooms, gentlemen's clubs and boardrooms of Philadelphia's WASP elite. An exercise of *noblesse oblige* as much as class self-interest, it harked back to a prewar model of incipient gentrification that could be found, for example, in New York's, Boston's or Washington's "best society." But neither Georgetown nor Beacon Hill nor Society Hill could remain the exclusive preserve of blue-book socialites. There were not, after all, enough Old Philadelphia WASP families to cover all of Society Hill, never mind gentrify the entire city, and so the process was quickly expanded to include the upper middle classes and professionals, whose breeding may have drawn sighs in some quarters but who at least had the money to do the job. By the 1970s, the patrician past may have lived on in a few chandeliered drawing rooms of those for whom gentrification was as much a lark as a civic duty, but by the 1970s real estate profits, tourist enticements and the *nouveaux riches* had taken over Society Hill. Patrician owners lived next door to well-heeled apartment renters, and by the 1980s an influx of yuppies had changed the tone of these neighborhoods entirely. The success of Society Hill had made its few quaint pubs and restaurants vibrant nightspots and the civic spirit of the 1970s was aimed less at abandoned buildings than at the Saturday night noise of visiting plebs.

Society Hill was in a second respect quite extraordinary. Although much gentrification in the US and in Britain has enjoyed public subsidy in one form or another, such strict orchestration of the process this early was rare. It has to be said, of course, that finance capital in particular was never far from the center of decision-making, and indeed organizations like GPM and the OPDC operated very much as pressure groups to manipulate local and federal initiatives in such a way that private-market operators would receive subsidies for rehabilitation and redevelopment while bearing very little of the risk. In institutional terms, Society Hill appears like a decentralized prototype for the kinds of larger downtown redevelopment schemes that mushroomed in the late 1970s and 1980s, from the Rouse arcades in Baltimore's Harborplace or San Francisco's Fisherman's Wharf (or indeed Philadelphia's Gallery) to Sydney's Darling Harbour or the London Docklands. Indeed, by the 1970s Society Hill built its own small commercial arcade. In different ways each of these cases combined similar coalitions of central and/or local government, public–private development corporations, business pressure groups and international development capital. But the parallel should not be overdrawn: the Rouse projects had become formulaic by the end of the 1970s while the Docklands involved an unprecedented scale of redevelopment and privatized control (A. Smith 1989; Crilley 1993), and it went bankrupt (see Fainstein 1994). By the other token, if Society Hill represents a more activist connection of state and private interests than was traditional in Europe until very recently, it actually stands in stark contrast to the experience of Amsterdam, where state regulation of the market significantly blocked large-scale gentrification, at least until the 1980s (see Chapter 8).

In Society Hill, as in gentrifying downtowns and inner cities in many countries, financial capital came into its own in the real estate markets of the early 1970s. Having assisted the devaluation of such neighborhoods in previous decades through redlining and the denial of housing finance, these increasingly globalized and diversified financial institutions, seeking alternatives to increasingly languid investments in the industrial, consumer and other sectors, poured large amounts of capital into real estate with an unprecedented amount directed toward central urban development. In the United States this put gentrification in more direct competition not only with the suburbs, but with the Sunbelt, Europe and the Third World. In Britain, by contrast, the competition for funds is less extensive insofar as mortgages generally come from a specialized sector of capital that has traditionally been more restricted from other financial dealings. Likewise, British gentrification began in the 1950s with very little state support and largely without the stimulus of such powerful private organizations as GPM; only later, with rising concern over inner-city disinvestment, did the central government offer a variety of improvement grant schemes (Hamnett 1973) to encourage gentrification. And only in the 1980s did large-scale schemes such as those associated with the docklands appear. Building societies had rarely withdrawn so completely from the inner-city housing market as their US counterparts. The more active involvement of the US state at an earlier stage of gentrification therefore speaks both to the more instrumental relationship between capital and the state in the US and to the depth of disinvestment.

In the US context, then, it makes sense to see Society Hill as a transitional project. Postwar urban renewal legislation – especially the 1949 and 1954 Housing Acts – was primarily aimed at the revitalization of central urban *economies* through housing reconstruction, and in this respect Society Hill was a unqualified success. It took such extraordinary levels of state control and subsidy to demonstrate the possibility of "renewal" in the first place and to absorb the economic risks involved. Despite the ambitions of legislators and developers, however, gentrification at this stage remained largely a specific set of processes in the housing market, enjoying more or less state support. But two things happened by the early 1970s to change this. In the first place, the well-publicized financial success of projects like Society Hill encouraged other developers to invest in rehabilitating old working-class neighborhoods with the benefit of less generous state subsidies and without such a blanket absorption of the risk. The rent gap, in other words, was coming to be exploited profitably enough through the private market.

But second, gentrification ceased to operate as a relatively isolated process in the housing market but was instead increasingly bound up with a broader urban restructuring that followed the political upheavals of the 1960s and the global economic depression of the early to mid-1970s. Not only housing, but employment patterns, social relations of gender and class, and the functional division of urban space were all being restructured, and gentrification became a part of this larger urban restructuring. This meant both a lubricated access to global (or at least nonlocal) capital as well as to a whole new demography of housing consumers. The gentility of Society Hill gentrification, to the extent that it had in the first place existed beyond the superlatives of boosterist accounts, was truly swamped. Instead, Society Hill became a popular magnet from which gentrification spread southward into South Philadelphia and Penn's Landing (Macdonald 1993) and westward toward Rittenhouse Square.

The appearance of success, then, is everywhere in and around Society Hill – from the neat propriety of the renovated facades to the equally neat propriety of the personal and corporate ledgers of those involved. The contradictoriness of this successful appearance cannot be found in the fine historic architecture of Society Hill or in the prestigious plaques that designate historic houses, all of which erase their own past. But it can be found in a few statistics about the area. By the 1980s Society Hill had doubled its population as compared with that in 1960; 63.8 percent of the adult population was college educated in 1980 compared with 3.8 percent in 1950; median family income in 1980 was over $41,000; the area boasted 253 percent of the median city income compared with 54 percent in 1950; and median house price had risen to $175,000. Owner-occupied housing in Society Hill was priced at more than seven times the city average (Beauregard 1990).

The contradictoriness of Society Hill's "success" can also be found buried deep in Redevelopment Authority files. In the first place, it was a quite qualified success for the city as a whole. Gentrification is widely justified as an enhance-ment of a city's tax base, a "triumph" that can potentially bring higher property tax returns and thereby enhance the "economic vigor" of the city (Sternlieb and Hughes 1983). And this was one of the main justifications given for Society Hill. In fact, since less than 20 percent of Society Hill's early residents actually returned from the suburbs, most coming from elsewhere in the city (see Table 3.1, p. 54), any increased property tax revenue owing to the area's gentrification therefore overestimates from the start its true contribution to city coffers since nearly 80 percent of gentrifiers already paid city taxes. Not that the increase was significant in any case. Total annual property tax from Society Hill was $600,000 in 1958 and only $1.7 million by 1975 (Old Philadelphia Development Corporation 1975). Since part of this increase was due to a higher city tax rate, the project resulted in an extra annual income of well below $1 million. The bulk of this remaining increase can be attributed to inflation over the seventeen years prior to 1975. By any reckoning, compared with the city's fiscal budget of $1.5 billion in the same year, the enhancement of the tax rolls was not substantial.

The low tax revenues in the area are most likely a result of low assessed values in Society Hill. While house prices increased by over 500 percent in Unit 1 between 1963 and 1975, the total assessed value of property in Society Hill did not even double; it rose from around $18 million in 1958 to approxi-mately $32 million in 1975. It is widely alleged that as a politically powerful community, Society Hill has succeeded in keeping its assessment values artificially depressed. The cost of Society Hill's success seems therefore to have brought a certain motionlessness to the bustle of urban renaissance. It has revitalized itself by bringing back those who already lived there; the city's benefits have just about matched its costs.

Second, Society Hill was successful, it is true, for its new residents, for its planners and for Alcoa, but, as the Redevelopment Authority files reveal, some 6,000 residents of Society Hill were displaced from 1959 onwards to facilitate the gentrification. To them, the success was surely far more qualified. Under the merest requirement from federal law, Redevelopment Authority personnel evicted residents, mostly tenants, on short notice and with derisory relocation assistance and compensation, if any at all. No good statistics were kept on the evicted population, of course, but they were disproportionately poor, white, black and Latino working-class.

Society Hill was indeed one of the projects that earned urban renewal its sarcastic reputation as "Negro removal." But the "whiting of Society Hill," as a local newspaper put it, did not happen without a struggle. Claiming that they and their ancestors had been in the neighborhood for more than a hundred years, African-American women led the fight against forced displacement. A group known as the "Octavia Hill Seven," named for the Quaker-dominated real estate society that was evicting them, formed an organization devoted to providing local housing for families displaced by the gentrification of Society Hill. This nonprofit organization was called the Benezet Corporation, after Anthony Benezet, a French abolitionist who formed the country's first free school for black children at Lombard and Sixth. It proposed that some vacant land near the edge of Society Hill be used to build housing for people with roots in the neighborhood who would be displaced by gentrification, and that the housing would be managed by the Benezet Corporation. By 1972, when this initiative was viscerally opposed, the original patricians had shrunk very much into the background of the ensuing gentrification struggles. A younger white neighborhood bourgeoisie insisted that the Benezet plan amounted to "public housing" by another name. "What I want to know," argued one recent immigrant to the neighborhood, "is by what authority do these people have roots? If you don't own, you don't have roots. What have they planted, their feet in the ground? I'll tell you this, we're going to fight this thing to the end" (Brown 1973).

The Octavia Hill Association, which owned the contested housing, offered to relocate the few remaining families in 1973 to West Philadelphia. "It gets me that white people see blacks as all alike," responded Dot Miller of the Octavia Hill Seven:

> [They] see nothing in plopping us down in a ghetto because they say, "You're black, you'll feel at home there," or something like that. Well I'll tell you I don't know how to live "black." I only know how to live period. This area has always been integrated and we were taught to see people as people. This is my home and I intend to stay.
>
> (Brown 1973)

Dot Miller and the Octavia Hill Seven were eventually evicted. "The market" was the major vehicle through which the white establishment fought to the end and won.

William Penn's "holy experiment" initiated on the same site 280 years earlier – "that an example may be set up to the nations" (Penn, quoted in Bronner 1962: 6) – billed itself not only as a "good thing" but as a new thing and as a necessary thing, much like the reincarnation of Society Hill after 1959. The darker side of its success, however, suggests that while the details of Society Hill's gentrification were exhilaratingly new for those who could profit from it, the losses involved tell a much older story:

The intimate connection between the pangs of hunger of the most indus-
trious layers of the working-class, and the extravagant consumption,
course or refined, of the rich, for which capitalist accumulation is the
basis, reveals itself only when the economic laws are known. It is otherwise with the "housing of the poor." Every unprejudiced observer
sees that the greater the centralization of the means of production,
the greater is the corresponding heaping together of the laborers,
within a given space; that therefore the swifter capitalistic accumulation, the more miserable are the dwellings of the working people.
"Improvements" of towns, accompanying the increase of wealth, by the
demolition of badly built quarters, the erection of palaces for banks,
warehouses, etc., the widening of streets for business traffic, for the
carriages of luxury, and for the introduction of tramways, etc., drive
away the poor into even worse and more crowded hiding places.

(Marx 1967 edn.: 657)

CATCH-22

The gentrification of Harlem?

Whites, they become urban pioneers. They're another variety of frontiersmen. They live this daring life and it's part of what they do. But then the outpost is right around the corner. They'll press a button and goddamned police'll be there just like that. . . . Of course they capture this sort of thing, park area, park space. There's good transportation routes, bus comes through here. They capture properties like this. What they intend for Harlem is that it not be Harlem again.
(Harold Wallace, Harlem resident)[1]

If gentrification began as a relatively isolated happening in the housing markets of a few select neighborhoods in the largest cities of North America, Europe and Australia, by the late 1970s (following a global economic recession) it had become an increasingly pervasive, trenchant and systematic occurrence. Reinvestment in urban middle-class residential rehabilitation and redevelopment had become more and more synchronized with a larger economic, political and social restructuring which, since the 1970s, has systematically altered the physical landscapes and cultural and economic geographies of cities up and down the urban hierarchy (Fainstein and Fainstein 1982; Kendig 1984; Williams 1984b; M. P. Smith 1984; N. Smith and Williams 1986; Beauregard 1989). Along with residential restructuring, this process often involves a partial recentralization of some professional, financial and producer services employment in new downtown office complexes; commercial revitalization (a "boutiquing" of central city neighborhoods); a dramatic expansion in upscale recreational and cultural facilities (restaurants, fern bars, art galleries, discos) as well as mixed-use urban spectacle projects from marinas to "tourist arcades" (Baltimore's Harborplace or London's St. Katharine's Dock, for instance).

That Harlem has become subject to gentrification at all speaks directly to the deepened significance of the process since the 1970s. Located in northern Manhattan in New York City, Harlem is a preeminent national and international symbol of black culture (Figure 7.1), and at first sight, perhaps, a highly unlikely target for a makeover. As the German magazine *Der Spiegel* put it, in the title of an article on the gentrification of Harlem: "Oh baby. Scheisse. Wie ist das gekommen?" ("Oh baby. Shit. How did that happen?") (Kruger 1985). The public representations of Harlem are manifold, intense, resonant, and highly imbricated with definitions of black identities. There is the Harlem of the Harlem Renaissance (Anderson 1982; Baker 1987; Bontemps 1972; Huggins 1971; Lewis 1981) or the Harlem of the 1960s – Malcolm X, Black

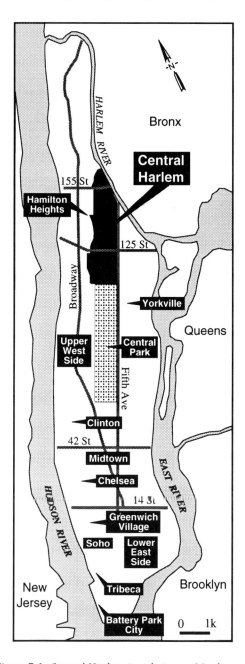

Figure 7.1 Central Harlem in relation to Manhattan

Power, the Black Panthers. But there is also Harlem the ghetto, the everyday Harlem for more than 100,000 people, predominantly poor, working-class and black; Harlem the community, the refuge from racism, starved for services; and there is Harlem the landscape of physical dilapidation, landlord criminality, social deprivation, street crime, police brutality, drugs. Harlem as haven; Harlem as hell (Taylor 1991). The latter is real enough for local residents even as its near-monopoly of media representations of Harlem magnifies racist stereotypes of variously threatening or exotic danger.

Constructed initially as a mixed middle- and working-class area in the last decades of the nineteenth century and located on the northern edge of Central Park, Harlem's housing stock generally comprises five- to six-storey tenements along the north–south avenues, studded with town houses and brownstones on the cross-town streets. Most of this housing stock was built in the major construction boom that followed the recession of 1873–1878 or in the later building phase at the turn of the century. The 1878 resumption of construction in an expanding Manhattan was funneled north and accompanied by the construction of an elevated railway into the recently annexed Harlem. Over the next fifteen years the lion's share of contemporary Harlem was built as a tony middle-class suburb to the downtown city to the south; construction was largely completed in a smaller boom from the late 1890s to 1904. The following year another recession set in and although it was less harrowing economically than those of the 1870s and 1890s, its effects in Harlem were worse: "The inevitable bust came in 1904–1905. Speculators sadly realized afterward that too many houses were constructed at one time" and the resulting glut led to widespread vacancies; "financial institutions no longer made loans to Harlem speculators and building-loan companies, and many foreclosed on their original mortgages." Much of Harlem "solemnly settled beneath a sea of depreciated values," concluded historian Gilbert Osofsky (1971: 90–91), quoting from *The New York Age*.

Faced with imminent ruin, numerous white landlords, owners and real estate companies took the unprecedented (and for them, desperate) move of opening up their recently built apartments and houses to black tenancy and ownership, sometimes through the intermediacy of black-owned real estate agents and companies. This came precisely at the time when traditional black neighborhoods like the Tenderloin were expanding dramatically beyond their established boundaries, given the beginnings of significant migration from the South. Blacks in any case were known to pay higher rents for lower amounts of space, and this led not only to the systematic subdivision of larger homes built with the white middle class in mind, but to a continuation of the disinvestment that had begun in 1905 (Osofsky 1971: 92). In a precursor to blockbusting, as it came to be known in the 1960s, white real estate agents and owners played on whites' racist fear of a "black influx," prompting them to sell at deflated prices, then raising the price for incoming black families on the grounds that they were moving into an exclusive middle-class suburb.

As the white middle class moved out to the suburbs and the black migration from the South accelerated during World War I, Harlem's population became increasingly black, and by the 1920s the Harlem Renaissance placed the area squarely at the forefront of black culture. New construction had effectively ceased by the beginning of World War I, however, and housing disinvestment deepened during the Depression. There would be little significant private reinvestment in Harlem until the 1980s except for undertakings that were partly or wholly funded by the state (largely in the 1950s and 1960s), and the population that concentrated in the area was overwhelmingly poor and working class. By the time that Harlem again made international headlines in the 1960s, it had been transformed into a slum and quickly became the most notorious symbol of black deprivation in the US.

In short, black residents – middle-class and working-class – who moved into Harlem in the early years of the century largely saved the financial hides of white landlords, speculators and builders who had overdeveloped. In turn, these residents, their children and their children's children were repaid by a bout of concerted disinvestment from Harlem housing that has lasted for nine decades.

Although the neighborhood's history of disinvestment and decline is in many ways typical of other neighborhoods facing gentrification, Harlem is quite atypical in other ways. Most important, Harlem is a black neighborhood; during the last two decades of the twentieth century, according to the census, only 4–7 percent of Central Harlem residents were nonblack. Gentrification in the US has certainly led to the displacement of black as well as other minority populations, but because many of the black urban neighborhoods had been targeted earlier by urban renewal, and because white middle-class gentrifiers have generally been less squeamish about moving into white working-class areas, the earliest neighborhoods affected by gentrification have been dis-proportionately white or at least mixed. With some exceptions, heavily black neighborhoods have been perceived as harder to gentrify. An obvious exception is Capitol Hill in Washington, DC (Gale 1977), which has under-gone gentrification since the mid-1960s. In Britain, one could point to Brixton which has a similar history of disinvestment and the ghettoization of a black British and Caribbean population, although Brixton always had a large white population too, unlike Harlem. In the 1980s Brixton began to transform, in the words of one writer, "from a riot-torn battleground to a gentrified playground" (Grant 1990).

Another important characteristic of Harlem is its size. Harlem is much larger than Capitol Hill or Brixton. Its total population is over 300,000 and it covers an area of about four square miles. Perceived by the middle class (espe-cially the white middle class) as highly threatening, having a universally depressed housing market, and possessing a cohesive social and political identity, Harlem represents a challenging obstacle for gentrification in New York City. Its location on the other hand – immediately north of Central Park "and only two stops from midtown on the A train" (Wiseman 1981) – does promise considerable economic opportunity for developers who initiate gentrification. With this much at stake, it is little wonder that on the one side Harlem has been seen as a supreme test for the gentrification process, while on the other gentrification is seen as a powerful threat to Harlem residents, who are dependent on the availability of housing at rents well below Manhattan market levels and on the availability of community support systems in lieu of private and public provision.

Harlem in the early 1980s, then, was susceptible to gentrification primarily because of two defining characteristics: on the one hand, its location close to one of the highest-rent districts in the world; on the other hand, despite this proximity, the neighborhood's sustained disinvestment throughout most of the twentieth century led to its having inordinately low rents and land values. The two and a half miles and two subway stops from midtown to Harlem represented by the 1980s one of the steepest imaginable rent gradients.

But Harlem's incipient gentrification also has to be placed in the context of wider developments. During the 1970s, New York City lost population, falling from a peak of nearly 8 million in 1971 to just over 7 million in 1980. Manhattan followed this trend, falling from 1.54 to 1.43 million in the decade, but in the same period the number of households in Manhattan actually increased by 2.5 percent (Stegman 1982). Along with this increase in households, the gentrification process, which had certainly been evident before, began to flourish in the late 1970s, especially following the depression of 1973–1975 and the simultaneous fiscal crisis that enveloped New York City. The displacement of economic crisis in the private as well as public sphere in the late 1970s led not only to reinvestment in residential development but to an unprecedented surge of new office construction perhaps best symbolized by the World Financial Center in Battery Park City. The restructuring of New York City's economy at precisely this time – as increasingly a financial and control center in the wider global economy – earned it and several other cities their reputation as "world cities" in the 1980s.

Thus despite the continued population loss at the city level during the 1970s, gentrification shows up strongly for the first time with census tract data from the 1980 census. Chall (1984) documents the process in New York City.[2] Geographically, gentrification concentrated in the southern and western parts of Manhattan: SoHo, Tribeca, the Lower East Side, Chelsea, Clinton and the Upper West Side all experienced considerable rehabilitation of old building stock (Figure 7.1). It also affected several neighborhoods in the outer boroughs, especially in Brooklyn, and in adjacent New Jersey towns such as Hoboken.

It is against this background of extensive rehabilitation in areas closer to midtown Manhattan, rapidly rising housing costs and rent levels, and an extremely low city-wide vacancy rate of about 2 percent at the beginning of the 1980s that the gentrification of Harlem came under discussion by residents, planners and city agencies. By the late 1970s, Harlem represented Manhattan's largest concentration of working-class residences with virtually no gentrification.

There are several purposes, then, to this chapter. First, apart from the fact that it offers a case study of an urban area with an international reputation – the gentrification of Harlem would indeed be an event of some significance – it also seeks to document the process virtually at its inception, providing a baseline against which future trends can be assessed. Partly out of disbelief that past trends would be reversed, most researchers have tended to study neighborhoods only after gentrification is an accomplished fact. Even if the process is truncated or halted, a study of its origins is useful. But second, this chapter has a broader purpose. It is meant to cast some light on the debates over the causes and significance of the process. There is little disagreement that Harlem represents a difficult target for gentrification; to the extent that it takes place, we should be more inclined to see the general process of gentrification as trenchant and long term. If it were temporary and small scale, why would developers and incoming residents make such long-term investments here rather than in neighborhoods perceived as socially and economically less risky? Finally, in Harlem the potential effects of the process on local residents are perhaps more visible than in many other neighborhoods, and this will also be a subject for discussion.

MAPPING THE ORIGINS OF HARLEM GENTRIFICATION

Harlem stretches for more than two miles north of Central Park in Manhattan. On the East Side it extends south to Ninety-sixth Street while on the West Side its southern border does not now stretch much below 125th Street. Generally, it includes Manhattan's Community Districts 10 and 11 and the northern part of Community District 9. During the late 1970s and early 1980s, some new construction and renovation began in the eastern section above Ninety-sixth Street, and there were also the beginnings of renovation in the western section, especially in Hamilton Heights and Sugar Hill. The section of West Harlem around Columbia University ("Morningside Heights") was a highly protected white enclave for most of the twentieth century. But the heart of Harlem lies in the central area directly north of 110th Street and Central Park (Figure 7.2). Unless this area of Central Harlem is gentrified, it is unlikely that the rehabilitation and new construction along the edges will amount to anything very significant. The earliest media reports of gentrification focused more on the eastern and western edges of Harlem, with far fewer reports of activity in the central area (Lee 1981; Daniels 1982; Hampson 1982). The core area of Central Harlem, covered by Community Board 10, is therefore the crucial battleground. This area stretches from 110th Street in the south to 155th Street in the north, and from Fifth Avenue in the east to Morningside and St. Nicholas Parks in the west.

Table 7.1 provides a statistical profile of this Central Harlem neighborhood in 1980, on the eve of gentrification. The comparison with Manhattan emphasizes the social, physical and economic contrast between Harlem and the rest of Manhattan. This statistical picture clearly reaffirms the popular perception of Central Harlem as predominantly poor, working-class and black; but it also shows that in the pregentrification decade of the 1970s, the population of Central Harlem declined by a third. Central Harlem does have a middle class, but it is very small; it is marked by a disproportionately low percentage of college graduates, and a small number of high-income households. Median rents are 25 percent lower than the Manhattan average, and 62 percent of the housing units are publicly owned, operated or assisted. A quarter of all housing units were abandoned, housing conditions are bad and the private housing market has historically been very soft. The contrast with the rest of Manhattan could hardly be more stark.

As other census figures reveal, while per capita income in the whole of Manhattan increased by 105.2 percent during the 1970s (with no correction

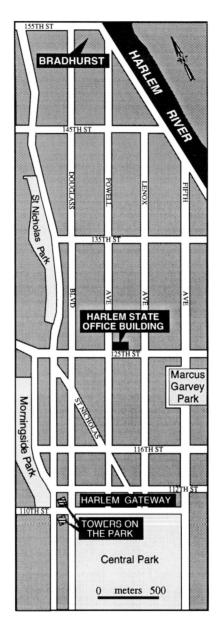

Figure 7.2 Central Harlem

for inflation) and increased by 96.5 percent throughout New York City, in Central Harlem the increase was only 77.8 percent, about 20 percentage points lower than the inflation rate for the decade. The decline of family income was even more marked. But not only did relative standards of living for Central Harlem residents drop during the 1970s; absolute standards of living did too. Thus as regards housing costs, median contract rent rose by 113 percent, out-stripping income by fully 35 percentage points. Manhattan and New York City

Table 7.1 Statistical profile of the Central Harlem (New York) population, 1980

	Central Harlem	Manhattan
Percentage population black	96.1	21.7
Per capita income ($)	4,308	10,992
Percentage high-income households ($50,000+)	0.5	8.4
Percentage low-income households (less than $10,000)	65.5	37.4
Percentage college graduates (adults with 4+ years of college)	5.2	33.2
Median contract rent ($ per month)	149	198
Percentage managerial, professional, and related occupations	15.9	41.7
Private property turnover rate per year (1980–1983) (%)	3.3	5.0
Population change, 1970–80 (%)	−33.6	−7.2
Percentage housing abandoned	24.2	5.3

Sources: US Department of Commerce 1972, 1983; City of New York, Department of City Planning 1981; Real Estate Board of New York 1985

median rent increases, by comparison, were higher than in Harlem (141 percent and 125 percent, respectively) (US Department of Commerce 1972, 1983). The 1970s, then, were marked by continued and severe decline in both the social economy and the property economy of Central Harlem.

Within this general picture, however, there is considerable social and geographical variation; the general trend of economic decline was not universal. A disaggregation of the data at the census tract level provides clear evidence of an opposite trend in some areas. Since gentrification involves a symbiotic change in social class and physical housing stock, a strong argument can be made that, given available indices from the US census, the most sensitive indicators of gentrification will involve a combination of income and rent data. In Central Harlem, despite the general picture of economic decline, nine of the twenty-seven tracts experienced per capita income increases higher than the city average. If we turn to indicators of rent, we find that in most of these same tracts rent increases were also generally above the local average, indicating a change in the housing market as well as a change in the social and economic composition of residents.

When these tracts of economic expansion are mapped they reveal a distinct spatial pattern. Such a pattern would be expected in the case of gentrification because the process tends to be tightly concentrated in specific blocks and neighborhoods, at least in the beginning. Figure 7.3 shows the distribution of census tracts with increases in per capita income that are above the city average. There are two corridors of more rapidly rising incomes, one on the western edge of the district, the other on the eastern edge. The same two corridors emerge from a map of above-average rent increases. But how do we know that this pattern results from gentrification and not from some other set of processes?

Plate 7.1 Restored Harlem Brownstones on Manhattan Avenue, 1985

After closer examination of the census tracts, it seems that while gentri-fication might be occurring in the western corridor, the idea of an eastern corridor of gentrification is not tenable. The eastern corridor, from 126th Street to 139th Street, largely comprises a low- and moderate-income urban renewal project (Lenox Terrace) as well as several blocks of severely deteriorated tenements and town houses. There is no obvious explanation for the above-average increases in income here, but it is at least possible that this part of Harlem is experiencing some spin-off effect from the concentration of office employment (since 1971) in the new Harlem State Office Building, immediately to the south on 125th Street. There were no visual signs of signifi-cant residential rehabilitation or redevelopment in the early 1980s, and so it makes sense to discount this area from a map of Harlem gentrification. The remaining tract in the eastern corridor, to the south of Marcus Garvey Park, may, however, be undergoing the beginnings of gentrification. Some well-publicized rehabilitations of town houses began here, and the area was targeted by the City of New York in its auction of City-owned properties. At best, however, the process is in its infancy.

In the western corridor, however, there is firmer evidence of the beginnings

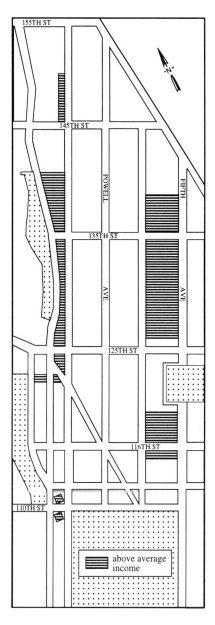

Figure 7.3 Above-average increase in per capita income in Harlem, 1970–1980

of gentrification. Above-average increases in income and rent, especially above 126th Street, are matched by an increase in the number of high-income households, yet the picture for professionals and college graduates is more ambiguous. This is particularly surprising because the western corridor borders upon the City College of New York, which would be expected to contribute graduates and "professionals" to the gentrification process. Still, the census data suggest the real possibility that gentrification had begun in this area by 1980. A more precise analysis demands that we examine a broader range of data, especially concerning the housing market.

An examination of housing market data for the early 1980s gives a more detailed picture of emerging gentrification. Between 1980 and 1984 there are ambiguous trends in the Central Harlem housing market, and these are shown in Figure 7.4, which graphs data on the volume and value of private residential sales (Real Estate Board of New York 1985). Although sales were generally flat coming out of the 1970s there is also a discernible decline in sales volume after 1981 in concert with the economic recession of the early 1980s, and it was followed by a decline of prices in 1983. There is little doubt that these declines represent local versions of national and international trends; nationally, residential sales volume declined 17.5 percent in 1982 over the previous year, and prices actually declined in many parts of the country for the first time in over a decade ("Home sales low . . . "1983). But a second important trend is also evident. Although the volume of sales did not pick up appreciably after the end of the recession in 1983, prices rose dramatically in 1984. This reflects a general perception by realtors, public officials and residents of the area that the market had heated up considerably, even if something of a wait-and-see attitude remained among potential investors: speculative investments increased after 1984 but at least in the beginning tended to involve smaller rather than larger investors (Douglas 1986).

As with the census results, these sales data for the area as a whole do not give the full picture. If we consider private residential turnover rates as an indicator of activity in the residential property market, then the overall rate of

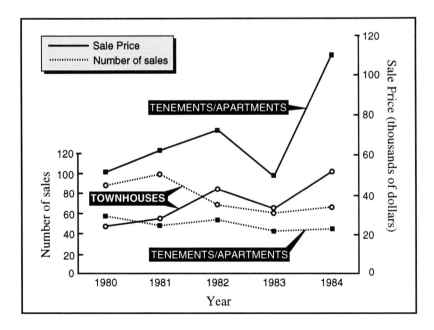

Figure 7.4 Volume and value of private residential sales in Harlem, 1980–1984

3.3 percent annually (Table 7.1) in the early 1980s suggests, as might be expected, a very slow property market in comparison with a 5 percent turnover rate for all of Manhattan. However, as Figure 7.5 shows, there is a highly uneven geographical distribution of turnover rates. The highest rate of private sales occurs in and around the same western corridor that emerged from the census data as a possible locus of gentrification. Further, it is apparent that in the most active areas, turnover rates are over 7 percent per year, appreciably more than even the Manhattan-wide average. This indication of increased activity in the real estate market concurs with earlier results. The Harlem Urban Development Corporation (HUDC) concluded as early as 1982 that in the triangle southwest of the St. Nicholas Avenue diagonal there had been a considerable increase in sales activity after 1978 (Harlem Urban Development Corporation 1982). A subsequent report reached a similar conclusion (AKRF 1982). The data through 1984 suggest a secular strengthening of this trend in the western corridor as a whole. Here the private residential market showed precisely the signs of intense activity associated with gentrification.

Although it is the western corridor that emerges as the area undergoing most significant changes in social composition and in the housing market, there are two other parts of Central Harlem where rehabilitation and redevelopment are beginning to take place. First, there is the area dubbed "Harlem Gateway" in the early 1980s, a bureaucratic naming which suggests vividly the aspirations and intentions of federal and local agencies for the area (Figure 7.2). Spanning the southern edge of Central Harlem, between 110th and 112th Streets, the major asset of the "Gateway" is that it hugs the northern edge of Central Park. This area was designated a Neighborhood Strategy Area by the Department of Housing and Urban Development (HUD) in 1979, meaning that it was targeted for HUD's major development programs, and it was also targeted by HUDC and various City agencies. By 1982, there were at least five Section 8 low- and moderate-income federal projects active in the area, involving the substantial rehabilitation of nearly 450 housing units. Although these projects themselves were hardly a spur to gentrification, they did signal local and federal government commitments to the "Gateway', and led to the announcement of several new projects requiring significant private investment. Very quickly this neighborhood was recognized as an area on the "verge of major redevelopment" (Daniels 1984). Most important are several condominium projects on and around Lenox Avenue and on the western edge of the "Gateway."

The largest and most significant development, however, has been the 599-unit "Towers on the Park" condominium. This was planned in the early 1980s, ground was broken in October 1985, and the building was opened in 1988. It has served as a southern anchor for the gentrification of Harlem. "Towers on the Park" was conceived and organized by the Rockefeller-inspired New York City Housing Partnership, which had developed a number of housing projects around the city that combine market-rate housing with some federally subsidized units. Construction was carried out by the Glick Organization, a major US urban development firm. The significance of Towers

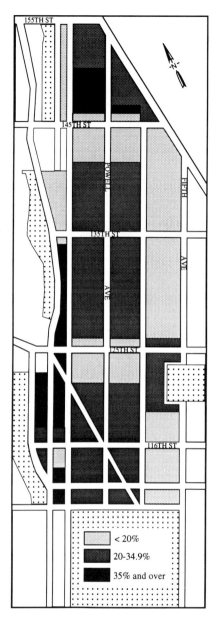

Figure 7.5 Private residential turnover rates in Harlem, 1980–1984

on the Park is far-reaching: it dramatically alters the physical, social and financial landscapes of Central Harlem. Physically, the condominium represents two twenty-storey towers (and several smaller ones) clustered around Frederick Douglass Circle on the southwestern edge of Central Harlem. These buildings dominate the horizon, overtowering everything else in the area except for the Harlem State Office Building a mile to the north. Financially, it

Plate 7.2 Towers on the Park, Harlem, New York

is an equally extraordinary undertaking. In the summer of 1985 the City received a $6 million Federal Urban Development Action Grant to subsidize the condominiums, and this triggered an unprecedented Chemical Bank loan of $47 million to finance construction (Oser 1985). This is by far the largest private residential capital investment in Harlem in decades; to put it in perspective, the Chemical Bank loan amounts to nearly eight times the total private mortgage financing that went into the whole of Central Harlem in 1982. In this one development can be seen both the severity of redlining in the past and the potential for gentrification in the future. The total projected cost of "Towers on the Park" exceeded $70 million.

In social terms, 20 percent of the apartments were earmarked for moderate-income buyers (1986 income less than $34,000); 70 percent for middle-income (1986 income between $34,000 and $48,000); with the remaining 10 percent available to high-income owners earning over $48,000. The cheapest apartments ranged from $69,000 to $110,000 while the most expensive were advertised at $340,000 (New York City Partnership 1987). By any standards, even the cheapest condominiums in this development were beyond the means of most residents of Harlem, where median per capita income was only a fraction of $34,000.

The second area of some activity since the early 1980s, outside the western corridor, is the area around Marcus Garvey Park. The census data on income and rent give a mixed picture, suggesting above-average increases in the tract to the south (Figure 7.2) but below-average increases immediately adjacent to the park. Most of the activity in this area began after 1982 when the City began its sealed-bid auction program with twelve brownstone properties that had been taken in property tax foreclosure proceedings. Three-quarters of these were in the immediate vicinity of Marcus Garvey Park. The properties were to be rehabilitated by those who won the auction, in what was seen as a trial run by the City. Although completed renovation of these properties following the auction was very slow,[3] the city administration was nonetheless determined to continue the auction program and to expand it. The area around Marcus Garvey Park remained a prominent focus in this program and was also highlighted in media publicity announcing the possibility of an imminent gentrification of Harlem (Daniels 1983b; Coombs 1982). Between January 1980 and June 1983, a total of thirty town houses were sold to private buyers in the tract adjacent to the park, the third highest total for Central Harlem. Other physical signs of gentrification became evident, from newly installed mahogany doors to sand-blasted facades.

This simultaneously rising property market and rising socioeconomic profile of the western corridor of Central Harlem especially constitute the hallmark of gentrification. Such a combination of shifts is unlikely to occur in Harlem for any other reason. The socioeconomic change indicates that the heating up of the property market is not simply the result of speculation, although the latter certainly occurs (Douglas 1986), most likely beginning with the early-1980s surge in property values. Conversely, the rising property market indicates that socioeconomic changes in the western corridor are tied to an upward revaluation of the physical structures. Also significant is the fact that in this western corridor there was no significant racial change by the early 1980s, no white influx. This suggests that the earliest rehabilitation represented a process of black gentrification. This indeed was the population that Taylor (1991) focused on in her attempt to emphasize the class dimensions of gentrification within Harlem.

Even in the western corridor, however, gentrification in the mid-1980s was sporadic and not at all generalized. To emphasize the preliminary character of the process, it is possible to compare sales data for Central Harlem with similar data for other clearly gentrifying areas in Manhattan. While Central Harlem had a total of 635 residential property transactions in the five-year period from 1980 to 1984 (for a total of $30 million and an average of $47,500), other clearly gentrifying areas of Manhattan experienced much greater levels of activity. Yorkville, on the eastern border of East Harlem and the Upper East Side, had 121 transactions in 1980 and 1981 for a total of $106.1 million and an average price of $877,000. Clinton, west of Eighth Avenue between

Forty-second and Fifty-Seventh Streets, had 142 sales in the same two-year period for a total of nearly $46 million and an average sale price of $322,000 (AKRF 1982). Although these data are not strictly comparable given different housing stock in different areas, the comparison does suggest that while the property market in parts of Central Harlem is beginning to show signs of gentrification, the phenomenon remains on a comparatively small scale. Further, it is important to remember that the 1970 base levels of the indicators used here (e.g. income, rent) are lower than the city average, as are property sale prices, and so large percentage increases, especially for the very small census tracts in the western corridor, do not necessarily mean large-scale activity.

Given that the core of Central Harlem represents some of the most deteriorated and devalued properties in the city, it is wholly to be expected that the process would begin at the margins. In some cases, such as the northern part of the western corridor, this might reasonably be considered the result of spillover from already gentrifying areas such as Hamilton Heights. Elsewhere, however, this is not the case: the Marcus Garvey Park area does not lie in close proximity to any other gentrifying area, and in the southern sector of the western corridor, the metamorphic schist outcrops of Morningside Park have been deployed as an effective barrier to social and economic intercourse between Harlem below and Columbia University's Morningside Heights above the hill. It can hardly, therefore, be considered spillover. Nonetheless, since it is less risky in market terms to attempt to level off the rent gradient at the margins, where higher land values act as an economic anchor, than to begin in the center, it is the edges that have attracted initial attention. As we shall see below, this is also the strategy of the City's Redevelopment Plan for Harlem.

MOMENTUM, DYNAMICS AND CONSTRAINTS

Perhaps the most important determinant of Central Harlem's future is local and national state policy. The City is particularly important because it is the major landlord in the neighborhood and because it embarked on a pro-gentrification strategy in the early 1980s. More than 60 percent of the housing units in Central Harlem are state-owned or assisted: in the early 1980s the City owned 35 percent of the housing stock (most of it taken in foreclosure proceedings) while another 26.4 percent was either public housing or was constructed with public assistance (Table 7.2). A further taking of foreclosed properties in the mid-1980s reduced the proportion of remaining private units closer to 30 percent. City strategy therefore looms large in the fate of Harlem, and the strategy has indeed been to encourage gentrification.

In the summer of 1982 the then mayor, Edward Koch, released copies of a "Redevelopment Strategy for Central Harlem," prepared by a special task force (City of New York, Harlem Task Force 1982). The release of this report galvanized the perception in Harlem that the neighborhood was indeed on the auction block (Daniels 1982). The Harlem task force called for a selective targeting of "stronger" anchor areas in Central Harlem in the attempt to induce a redevelopment that is "economically integrated" (p. 2). The Redevelopment

Table 7.2 Ownership of housing units in Central Harlem, New York, 1983

Ownership	Housing units	%
Public housing	8,144	14.6
City-owned housing*	19,588	35.2
Publicly assisted private housing		
Mitchell-Lama	2,520	4.5
Federal Title I	3,528	6.4
Urban Development Corp.	501	0.9
Private	21,399	38.4
Total	55,680	100.0

Source: City of New York, Department of City Planning 1983.
*Buildings taken by city through *in rem* process

Strategy begins from the assumption that "with drastic reductions in Federal housing and economic aid," the emphasis would have to shift toward private-market investment and public–private partnerships: the "private sector . . . would have to play a pivotal role" (City of New York, Harlem Task Force 1982: i–ii). The overarching strategy, then, was to apply limited public funds to bolster areas where the private market was already becoming active (the western corridor, essentially), and to use anchor areas to the south (the Gateway) and the north (the stretch from Hamilton Heights to the middle-class enclave of Strivers Row around 138th and 139th Streets, where disinvestment has been less marked and private lenders still operate) in order to encircle the heart of Harlem. The City's target areas are shown in Figure 7.6.

A central plank of City strategy in Harlem in the 1980s involved the auction of City-owned properties. In a "dry run" several months prior to the release of the Redevelopment Strategy, the City auctioned twelve town houses to bidders who contracted with the City to rehabilitate them. Despite emerging opposition and widespread apprehension in Harlem (Daniels 1983a), the City government was ready by 1985 to implement a full-scale auction. One hundred and forty-nine additional town houses were put on the block: 1,257 bids were received; the winners paid between $2,000 and $163,000 for the properties; the average auction price was close to $50,000, and of the winners, ninety-eight were by prior agreement residents of Community Districts 10 (Central Harlem) or Community Board 9 (Morningside Heights and West Harlem and Hamilton Heights) (Douglas 1985). Perhaps most significant is that City officials succeeded, with the help of a $6 million grant, in persuading the Freedom National Bank, a black-owned bank with a strong commitment to Harlem, to provide purchase and renovation loans to auction winners at well below market rates (7.5 percent).

Mortgage data from the early 1980s suggest just how severely Harlem was redlined prior to this and other commitments. Of the $12 million invested in Central Harlem mortgages in 1982 (nearly all of which was for large multi-family dwellings) HUD provided 47.5 percent for six separate buildings. Most of the remaining investment (a further 34.5 percent) comprised "purchase money mortgages," that is, seller-financed mortgages. There were over thirty

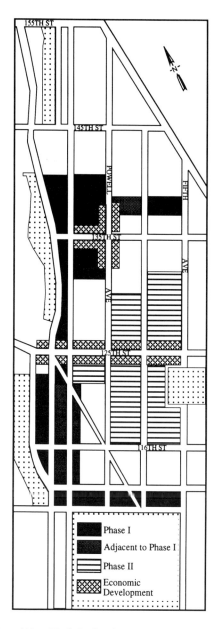

Figure 7.6 City of New York Redevelopment Strategy areas for Harlem

private institutional lenders in the market, mostly small local lenders, but no one of them accounted for more than 2 percent of the total mortgage money. That is, no single private financial institution ventured as much as $240,000 in the entire area in 1982 (City of New York, Commission on Human Rights 1983). Little wonder then that City officials measured their success largely in terms of their ability to attract the Freedom National and Chemical Banks as well as other sources of private capital to finance rehabilitation and new construction. These agreements signaled widely in the mid-1980s that private capital was beginning to perceive Harlem as a viable, even lucrative, investment for the second half of the 1980s – especially when backed by public funds. "With gentrification sweeping many parts of the city, investors and institutions were eager to lend to real estate buyers," the *New York Times* noted. "You had people fighting to give you money because the market was hot," said Ira Kellman, a Harlem property owner (Purdy and Kennedy 1995).

All of this activity clearly had an effect. In the 1980s, population loss in Harlem was dramatically slowed to less than 6 percent, and the nonblack population almost doubled, accounting by 1990 for 7.5 percent. Not unexpectedly, perhaps, given the dramatic expansion of high-paid professional jobs in the 1980s, per capita income in Harlem more than doubled, but it did not increase as rapidly as in the rest of Manhattan. And yet the high-income population of Harlem (measured as those earning over $75,000 in 1989) has increased much more rapidly than that of Manhattan as a whole. If this is a small absolute number – 3 percent of Harlem households compared with 19.4 percent for the borough as a whole – it is a significant one nonetheless (US Department of Commerce 1993). In the 1980s, as in the prior decade, the highest increases in income and rent levels are concentrated in the western corridor, the southern gateway area, and in the vicinity of Marcus Garvey Park. The most spectacular increase comes, not surprisingly, in the southwest corner of Harlem, where the opening of Towers on the Park has contributed to an almost 400 percent increase in the per capita income of census tract 197.02. In that tract, the per capita income for 1989 jumped to $18,399, nearly $4,000 higher than for any other tract in Central Harlem.

If gentrification in Harlem had been sufficient in the 1980s to show up in 1990 census data, the story was only beginning. The stock market crash of 1987, felt most intensely in the brokerages a few subway stops south of Harlem, induced by 1989 a virtually global economic depression. Property markets were severely depressed, and gentrification activity, which in many places survived earlier recessions unscathed (Badcock 1989; Ley 1992), was widely affected by this economic decline that ushered in the 1990s. In Harlem, work on many of the auctioned properties slowed or halted altogether, and in many cases (even prior to the depression) the original bidders were financially unable to carry out the rehabilitation, forcing the City's Department of Housing Preservation and Development (HPD) to contract with other buyers. Other private investments also dried up. Freedom National Bank went bankrupt in 1990, depriving the area of its most consistent source of mortgage funding, and other banks which had gingerly entered the Harlem market abruptly left. The disinvestment that had broadly characterized the area's real estate into the early 1980s descended again. Instead of optimistic stories about Harlem as the next frontier for the black bourgeoisie, local newspapers began a steady diet of remorse about a market gone bad, heroic but defeated landlords, declining rent roles, maintenance undone, decrepit and abandoned buildings (see for example Martin 1993).

The destruction wrought by disinvestment was tragically symbolized in

March 1995 when a building on 140th Street collapsed, killing three tenants and injuring seven. The building's owners, Marcus I. Lehmann and Morris Wolfson, were young, white professional landlords who had bought six Harlem buildings in the mid-1980s for $3 million. Although they borrowed $625,000 in 1987 and 1988, reputedly for repairs, the building steadily deteriorated. In 1991 Lehmann and Wolfson declared bankruptcy, claiming $12.7 million in debt, and following the building collapse it was revealed that 326 housing violations had been issued against the building. Fifty-one other cases have been brought by HPD against the landlords or their agents. In 1991, when a tenant sued on the grounds that the building's condition had caused injury to a child, the landlords claimed to be unable to afford insurance and offered instead to give the family free rent (Purdy and Kennedy 1995).

There is little doubt that the gentrification which began in the early 1980s was severely curtailed by the early 1990s, but it would be a mistake to conclude either that it shut down entirely or that the depression marked a definitive end to gentrification in Harlem (see Chapter 10). Even amid the depths of disinvestment, some old projects were completed (Oser 1994) and new ones begun. In 1992, for example, the New York Landmarks Conservancy began the rehabilitation of Astor Row, a row of small, set back, three-storey row houses with wooden porches on West 130th Street, originally constructed by William Astor between 1880 and 1883. Although some of the existing tenants and owners will remain in the buildings, previously vacant houses will receive new occupants, and the project is likely to anchor gentrification in the putative eastern corridor of Central Harlem above 125th Street. Plans for an International Trade Center and Hotel on 125th Street at Lenox Avenue, first vaguely posited as part of the economic development platform of the 1982 Redevelopment Strategy, were relentlessly pushed by the City, and in 1994 street peddlers and merchants were forcibly removed from the prospective site and from the sidewalks of 125th Street. Substantial low- and moderate-income housing development – part of the City's ten-year housing plan initiated in 1986 – has helped to recompose neighborhoods like Bradhurst in the north (Bernstein 1994).

CLASS, RACE AND SPACE

Since gentrification is only in its earliest stages in Central Harlem, the anticipation of change probably outstrips the reality. Nonetheless, significant changes are already apparent, the depression of the early 1990s notwithstanding (see also Badcock 1993), and this has crystallized a variety of social questions concerning gentrification in Harlem. Reflecting on the effects of the first wave of gentrification on Harlem, sociologist Monique Taylor suggests that a certain "crisis of identity" has emerged, especially for middle-class Harlemites (Taylor 1991: 113). On the one hand, Harlem is embraced as the home of black America, yet on the other hand, significant class differences separate those professionals who did move in during the 1980s from longer-term residents. Taylor provides vivid testimony of the contradictory connectedness of race and class identity resulting from gentrification. The new middle-class arrivees she talked with have a wide array of aspirations for the area, some nostalgic and romantic, others activist and practical. Some feel that defending Harlem as a place where black people can be insulated from the racism of the outside world – the "downtown" work world in particular – is the first priority. A number of women especially feel that a moderate influx of white middle-class gentrifiers would help to attract much-needed services to the area, while other new arrivees feel it would guarantee more attention from the City and help buttress property values.

The crux of Harlem's future lies with this interconnection of race and class. From the beginning, as the epigraph at the start of this chapter suggests, there was apprehension about what gentrification would bring. So far, although there are no precise figures, it is clear that despite prominent press reports featuring individual white gentrifiers in Harlem (Coombs 1982), the vast majority of people involved in rehabilitation and redevelopment in Central Harlem are African-American. It is true that 1990 census data register a significant increase in white householders in the area, especially in Towers on the Park, but elsewhere white middle-class migrants are a very small minority. Of the 2,500 applications received for the first round of the City sealed-bid auction, approximately 80 percent were estimated to come from African-American applicants.[4] At the same time, the City Redevelopment Strategy and especially the Harlem Urban Development Corporation have strenuously insisted that the "redevelopment" of Harlem is aimed to benefit Harlem residents themselves: poor and working-class blacks. How likely is this outcome?

In the 1982 auction, the City required that each entrant earn at least $20,000 per year (P. Douglas 1983), but in light of the difficulties experienced with that auction the 1985 auction was open only to households (or pairs of related households) with substantially higher incomes. Rehabilitation costs in the mid-1980s were estimated to be more than $135,000 for a medium-sized town house, and this required a minimum annual household income of between $50,000 and $87,500 for potential renovators ("Profile of a winning sealed bidder" 1985).

The 1980 census data reveal that only 262 households in Central Harlem had incomes of more than $50,000. In the whole of Manhattan, the number of black households earning more than $50,000 did not exceed 1,800; in all New York City there were fewer than 8,000 such families, implying that any African-American remake of Harlem would be disproportionately dependent on non-New Yorkers. Even from the beginning, then, whatever the rhetoric, it was clear that the gentrification of Harlem would not proceed far if it were simply a process of "incumbent upgrading" by Harlem residents. As much was suggested by E. M. Green Associates in a 1981 marketing study for a co-op building in the Gateway area (AKRF 1982). The study identified black households and individuals outside Harlem, indeed throughout the metropolitan area, as a potential market, a population that could be attracted to the area by the allure of living in Harlem, returning to one's roots. But by the 1990s the costs of renovation had doubled, and with them the required income for a viable renovation, further limiting the potential pool.

It is certainly possible that the economic vacuum in Central Harlem could be filled by non-Harlem blacks, and Taylor's (1991) research identified a number of households who moved "home to Harlem" (to use Claude McKay's (1928) title) in the 1980s. Yet if this group is to be the mainstay of Harlem gentrification, it would defy existing empirical research which suggests overwhelmingly that few gentrifiers actually return from the suburbs (see Chapter 3). Should Central Harlem follow this established trend, its major reservoir of potential gentrifiers will be New York City residents. The inescapable conclusion is that unless Harlem defies all the empirical trends, the process might well begin as black gentrification, but any wholesale rehabilitation of Central Harlem properties would necessarily involve a considerable influx of middle- and upper-class whites.

Although such an option is rarely if ever admitted in the public rhetoric supporting the redevelopment of Harlem, it is widely understood by Harlem residents and has been since the earliest intimation of gentrification (Lee 1981; Daniels 1982). It is significant, then, that in their prospectus and announcements for the Towers on the Park, the New York City Partnership entirely omitted any reference to the building's Harlem home, presumably in an effort to attract middle-class whites whose own identities could not square with living in Harlem. Towers on the Park was instead "located on the northwest corner of Central Park." And insofar as this development has made census tract 197.02 the whitest tract in Harlem, it has spearheaded a white gentrification alongside the black gentrification of Harlem.

Whether whites will be willing to move into Harlem in greater numbers remains unclear, and the larger question of finding a ready reservoir of housing consumers certainly constitutes a significant constraint on gentrification activity. But there are other constraints. The continuance of public financing is by no means certain in the context of even worse federal and local budget crises in the 1990s: the Harlem Urban Development Corporation was disbanded in 1995. And the reinvestment of significant amounts of private capital following the economic depression of the early 1980s is also uncertain. Equally uncertain is the response by poorer, long-term residents should gentrification proceed.

CATCH-22

The constraints on gentrification in Central Harlem, then, are considerable, but not necessarily insurmountable. There are also strong forces pushing for the redevelopment and rehabilitation of the neighborhood's housing stock. In the first place, there are the obvious assets of Central Harlem's location and transportation access. As professional, managerial and administrative employment continues to expand in Manhattan, as the number of households increases, and as the housing market becomes tighter, Harlem appears as an increasingly attractive candidate for gentrification. But despite its substantially underpriced housing in relation to the rest of Manhattan and the economic opportunity this represents, there is no automatic transformation of Harlem into a gentrified "haven." In locational and economic terms, there is no doubt that the potential for gentrification is there; the question is whether these economic and locational forces are powerful enough to overcome the constraints.

The fact that there are signs of the process in Harlem at all reaffirms the contention that far from being a curious anomaly, gentrification represents a trenchant and geographically extensive restructuring of urban space. In the case of Harlem, more visibly than in many other places, it is clear that the process involves "collective social actors" more than heroic individuals. In this case, it is not private capital alone that has played the leading role. Not until 1985 did a large potential influx of private mortgage capital begin to materialize. Led by City agencies, the state – in a number of institutional guises – has been most heavily involved in facilitating an upward momentum in the housing market.

The City's Redevelopment Strategy proposed to benefit Central Harlem residents, avoid large-scale gentrification, and produce an "economically integrated" community. It explicitly states that this "can be achieved without displacing the present residents of Harlem" (City of New York, Harlem Task Force 1982: 1, 2). In one very real sense, this is more possible in Central Harlem than elsewhere. The City owns such a vast stock of abandoned buildings (many of them vacant) and undeveloped land that it is possible for substantial rehabilitation and redevelopment to occur before low-income residents are directly threatened with displacement. The City has already begun rehabilitation of 6,000 units for mixed-income occupation (Bernstein 1994). But for the remainder of the City's privately based redevelopment strategy to succeed, two prerequisites are crucial. First, Central Harlem will have to attract a large number of outside residents. At first most of the new residents may be black, but as momentum builds many of them will necessarily be white. Second, the area will have to attract much larger quantities of private financing. Should these prerequisites be achieved, Central Harlem could be transformed from a depressed island of disinvestment into a "hot spot" of reinvestment, and integrated more fully into the Manhattan housing market. This would ultimately mean that large numbers of community residents would face displacement. Thus for Harlem, as for many other areas that have undergone gentrification, "economic integration" may be an impossible hope and "a little gentrification" may be too unstable a state to survive for long. And the City Redevelopment Strategy suggests as much: in Harlem, "economic integration" means bringing in wealthy people; "social balance" means an influx of whites.

Gentrification may be only one aspect of a larger urban restructuring that will fundamentally alter the face of Harlem along with other neighborhoods. In this context, Harold Rose has presented a bleak vision on the future of black working-class neighborhoods. "If the evolving spatial pattern of black residential development is not significantly altered," Rose says (1982: 139), the next generation of "ghetto centers will essentially be confined to a selected set of suburban ring communities located in metropolitan areas where the central city black population already numbers more than one-quarter million." We might add the corollary that if the evolving spatial pattern of gentrification in the central city continues, then not only will inner suburban ghettos burgeon, but the inner-city ones will shrink at the hands of white middle-class migrants.

This should not be taken as a prediction that Central Harlem will inevitably become a majority white neighborhood. We are barely embarked on such a course yet, even if there are signs of such a trajectory. Apart from the City's policy toward the area, there are two other major determinants of Central Harlem's future: the condition of the national and New York City housing markets, and the effectiveness of political opposition. If the housing market rebounds in the late 1990s then gentrification has a better chance of increasing its momentum. Further economic recession or collapse in the housing market generally would obviously be detrimental. But the second determinant of future gentrification is potentially the level of opposition from within the community, and this is where the congeries of race and class become important. There is a significant class division over different households' interests in gentrification. To the extent that the process again begins to drive up housing prices and even cause displacement, this division of interests between poorer renters and better-off owners will become clearer. Opposition to remaking Harlem for outsiders could well reshape or blunt the kind of gentrification that takes place. Alternatively, the City's major redevelopment of vacant buildings for mixed use could discourage further significant investment in private rehabilitation; without an across-the-board enhancement of employment and services, this latter scenario would likely lead to a new bout of disinvestment and the continued ghettoization of Harlem.

And yet Harlem since the 1980s has also experienced a modest "cultural revival" that could only fuel gentrification. The Apollo Theater was reopened and a new Multi-Media Arts Center was constructed on the site of the Renaissance Ballroom ("$14.5 Million Arts Project . . . " 1984). A small contingent of the City's white middle class has "discovered" several of Harlem's restaurants and clubs, and a larger white influx attends the "Harlem Week" festival every August. Bus tours around Harlem attract thousands of European and Japanese tourists; and Nelson Mandela's pilgrimage to Harlem shortly after he was released from his South African jail cell was widely and jubilantly celebrated by tens of thousands of people in the streets as a triumphant return by a long-lost son of Harlem. All of this in different ways may unwittingly lubricate the gentrification process as significant numbers of whites and middle-class blacks become more interested and intrigued by Harlem. Catching this spirit has been the aptly named *Harlem Entrepreneur Portfolio*, which promotes itself as "Harlem's newest brownstone newsletter': "The joys of living in Harlem are endless. The main one being a sense of community" ("Profiles in Brownstone Living" 1985).

It is difficult to avoid the conclusion that for Central Harlem residents, gentrification is a "Catch-22." Without private rehabilitation and redevelopment, the neighborhood's housing stock will remain severely dilapidated; with it, a large number of Central Harlem residents will ultimately be displaced and will not benefit from the better and more expensive housing. They will be victims rather than beneficiaries of gentrification. At present, there are no

plans for this contingency, either in the City's Redevelopment Strategy or elsewhere; none of the development strategies for Central Harlem even admit the
likelihood of displacement.

"CIRCLING THE WAGONS AROUND"

For more than ten years, before the corporation was disbanded, Dennis
Cogsville was president of the Harlem Urban Development Corporation, an off-
shoot of the defunct State Urban Development Corporation and a major
vehicle for gentrifying Harlem. Its offices are on the eighteenth floor of the
Harlem State Office Building, which provides a breathtaking vista of Harlem
stretching south to Central Park and the spires of midtown beyond. Cogsville
commuted in from New Jersey. He was a major player in launching "Harlem
Gateway" on the northern edge of the Park in the mid-1980s. "That's going
to be a tough project," he said, looking out at the tenements below:

> But here's how we're going to do it. Starting from 110th Street, we will
> make a first beachhead on 112th Street. You know, some anchor con-
> dominium conversions. Then a second beachhead up on 116th Street.
> That'll be a hell of a job. There's drugs, crime, everything up there. But
> we're going to do it. Essentially the plan is to circle the wagons around
> and move into Central Harlem from the outskirts.[5]

On 110th Street, sixteen months later, it was time for the groundbreaking
for "Towers on the Park." Dennis Cogsville was there, but it was US Senator
Alfonse D'Amato who took center stage at the ceremony. D'Amato, a powerful politician who would soon come under intense suspicion in relation to his
brother's corrupt real estate dealings, offered his own vision of how to make
Harlem "not be Harlem again." But he was confronted by an organized
community protest chanting and singing their opposition to the gentrification
of Harlem. Calling the condominium project "beautiful" and "New York at
its best," D'Amato glared at the protestors and, as the *New York Times*
described the scene, he then bellowed: "*I'd* like to sing too." He "broke into a
brief, off-key aria: 'Gen-tri-fi-ca-tion. Hous-ing for work-ing people. A-men'"
("Disharmony and housing" 1985).

ON GENERALITIES AND EXCEPTIONS

Three European cities

It has been suggested that while gentrification was first identified as such in Europe, specifically in London (Glass 1964), gentrification theory has been disproportionately based on the US experience, and European gentrification may not fit the theoretical arguments quite so neatly. The US itself may be an exception. By way of corollary, it has also been suggested that it may not be possible even to derive useful generalizations concerning the different experiences of gentrification in different cities and different neighborhoods. Local specificities overwhelm any possible generalizations. Since the preceding chapters have focused primarily on the US experience, in this chapter we shall look at the experience of gentrification in three European cities. The overarching question concerns the extent to which transcontinental generalizations are apt, indeed the extent to which any generalizations can be made concerning gentrification.

While Lees and Bondi (1995) make the most direct attack on supralocal generalizations, circumscribing significantly the work that theory can do, perhaps Musterd and van Weesep (1991) most explicitly entertain the plausibility of a distinctly European experience of gentrification, distinct that is from the US experience. While no simple dichotomy pertains, according to Musterd and van Weesep, they hold that a sufficient "Atlantic gap," to use Lees' (1994) term, does persist. As others have argued: "The process of gentrification in the US seems to be quite different from gentrification in the Western European metropolises, especially where social democratic governments control housing policy (e.g. Amsterdam, Stockholm)" (Hegedüs and Tosics 1991; see also Dangschat 1988, 1991).

A number of issues are generally raised in this context: the longer history of monetized production relations *vis-à-vis* the built environment in Europe; the shallower levels of disinvestment in European cities; the more *laissez-faire* involvement of the state in the urban land and housing markets in the US (whence so much of the gentrification literature has emanated); radically different histories of racial differentiation and homogeneity; and different cultural economies of consumption. Together these issues are often raised in support of the notion that European gentrification is systematically distinct from that of the US, or for that matter Canada or Australia.

With this argument in mind, this chapter will consider the experience of gentrification in Amsterdam, Budapest and Paris. The choice of these cities is guided by numerous considerations: an adequate extent of gentrification; availability of information; different experiences of gentrification; different types and levels of state involvement; and, in the case of Budapest especially, a quite idiosyncratic relationship to global capital and economic restructuring.

AMSTERDAM: SQUATTERS AND THE STATE

In retrospect, the "Battle of Waterlooplein" in 1981, and the simultaneous apex of the squatters' movement, may well mark the social watershed between two distinct periods in the urban history of late-twentieth-century Amsterdam. On the southeastern edge of the old sixteenth- and seventeenth-century city, Waterlooplein had been a symbolic public space for centuries: a wide, irregularly shaped square, it was a center of working-class Jewish life for several centuries, a primary target for early slum and shanty removal by the late nineteenth century, the site of a major market, and the center of the ghetto during the Nazi occupation. The area around the Waterlooplein was scavenged for wood toward the end of the war and in subsequent years experienced further deep disinvestment and physical dilapidation. After the war the square hosted a major flea market. Postwar plans to rebuild the square as a monument to the destroyed Jewish population never came to fruition, and by the late 1970s the municipal government announced plans to clear the square and its surrounds for the construction of a new Stadhuis (City Hall) and Opera House – dubbed Stopera. Coming on the heels of the earlier violently contested reconstruction of another neighborhood (Nieuwmarkt), in part to build the Amsterdam metro, the City's plans for Waterlooplein galvanized broad local opposition from residents, housing and antigentrification activists, squatters, feminists and poor people for whom the flea market was both an economic necessity and a way of life. Resistance to the high modern design of the Stopera, the largest redevelopment project in the city, coalesced in 1980 and 1981 in a series of violent police confrontations with demonstrators determined to prevent clearance in and around the square (Kraaivanger 1981). Particularly galling for those resisting the project was the plan to install what was seen as an elite opera house rather than some kind of public perform-ance space, a renovated people's market or much-needed housing. The "Stopera" was delayed but finally built, opening in 1986.

The story of gentrification in Amsterdam both predates and postdates this widely publicized clash, and, as with the fate of the Waterlooplein, it centres on the pervasive involvement of the state in the city's land and housing markets. While gentrification first emerged in Amsterdam in the 1970s, it has been fueled since the early 1980s by significant shifts in municipal housing policy. Questions of housing, gentrification, squatting and redevelopment have in fact been at the heart of municipal politics in Amsterdam, perhaps even more than in most cities, and they have played a central role in "the restructuring of Dutch cities" (Jobse 1987).

Like that of many cities in the advanced capitalist world, Amsterdam's modern decentralization planning was well-established by the early 1970s. Sufficiently successful was this policy, in conjunction with spatial shifts in employment and services, that the city population peaked in 1964 and fell by nearly 200,000 people in the next two decades; its 1980 population was estimated at 676,000 (Cortie *et al.* 1989: 218). During this period, urban development in the city was strictly controlled by the municipal government. A general plan guided physical construction in the city and national government subsidies regulated private-market construction. A regulated housing supply was thereby matched by regulation of demand: a strong system of allocation procedures and goals, centrally administered rent control, tenure protection for tenants, and homeowner tax deductions. As Dieleman and van Weesep argue, "few national governments, even among those of the social democracies of Western Europe, intervene so extensively in the housing market as that of the Netherlands." Its involvement is "virtually comprehensive" (Dieleman and van Weesep 1986: 310).

Urban policy in this period was geared largely toward the replacement of dilapidated neighborhoods and, somewhat in response to the resulting political opposition in the 1970s, toward the construction of social housing. Population loss, however, together with a significant suburbanization of office and other central urban functions as well as deindustrialization and the loss or decentralization of traditional urban industrial jobs, all threatened to empty Dutch city centres, and this led by the end of the 1970s to a significant shift in national and municipal housing policy. The new national urban policy of the 1980s valued a "compact city" and emphasized the recentralization of certain residential, professional, tourist and other service facilities and activities, especially in Amsterdam. To this end, vacant buildings and sites and structures in need of rehabilitation were identified by the municipal government; private developers were encouraged to carry out the renovation and provide housing, much of it on the private market. The new urban policy explicitly tried to reverse the decentralization and decline that had dominated the 1970s (van Weesep 1988; Musterd 1989), and indeed by the late 1980s the population had again risen above 700,000.

The cautious deregulation of the early 1980s was dramatically accelerated by the 1989 Housing Policy which, while maintaining a commitment to the provision of some social housing, effectively dismantled the pervasive system of subsidies and regulation (van Kempen and van Weesep 1993). This new "bias toward the private sector is manifest in the construction program" guided and subsidized by the state (van Weesep and Wiegersma 1991). Deregulation and privatization of the housing sector was driven as much by mounting budgetary constraints as by an ideological agenda, unlike in Britain perhaps, where Thatcher's privatization of housing was before anything ideologically led. It should also be noted that the privatization of social housing in the Netherlands applies predominantly to new construction and has not resulted in the kind of massive social housing sell-off that characterized the British experience in the 1980s. And while the recrudescence of a powerful momentum to the private housing market is undeniable in the 1990s, the private sector is nowhere as dominant as it has traditionally been in many other places, especially US cities.

The gentrification of Amsterdam over the last twenty or more years has to be understood in this context. Although conclusive empirical research has not yet been carried out, it is fair to conclude that by the early 1970s, when the first signs of gentrification emerged, a significant rent gap existed (Cortie and van de Ven 1981). As ever, the origins of this rent gap are not entirely straightforward. In the first place, highly inclusive state regulation of the land and housing market meant that Amsterdam never experienced the levels of private market disinvestment usually associated with US cities. Vacancy rates certainly rose and undermaintenance of older buildings (as well as more recent tenements and estates) certainly occurred but since maintenance was often subsidized by the government, the level of disrepair was not extreme. The abandonment of buildings was a distinct oddity, where it happened at all. Commercial buildings may well have been abandoned in terms of current use, but the title to the properties was never relinquished. In the second place, however, at the same time as it inhibited the development of a rent gap, state regulation in the shape of rent control simultaneously encouraged it by depressing broadly the rents and therefore sale prices of inner-city housing.

Given these controls, much of the gentrification of central Amsterdam involved a shift in tenure form – from nonresidential to residential, from partly subsidized to wholly private stock, from rental to ownership, or into a rapidly expanding condominium sector (van Weesep 1984, 1986).[1] Many neighborhoods therefore also experienced what Hamnett and Randolph (1986) have called a "value gap" (see also Clark 1991a). But as van Weesep and Wiegersma note, the expression of both rent and value gaps was significantly suppressed by the strictness of state regulation. "Although this system is still in place," deregulation since the 1980s means that "both a rent gap and a value gap occur where redevelopment is permitted" (van Weesep and Wiegersma 1991).

In their analysis of building rehabilitation permits, Cortie et al. (1989) demonstrate a remarkably concentrated pattern of private reinvestment in the historic center of the city with no parallel concentration in the suburbs. Further evidence suggests a similar but milder concentration of higher-income and professional households in the center (Cortie et al. 1989; Musterd and van de Ven 1991). In fact gentrification has taken root in three main areas: the Old City; the seventeenth- and eighteenth-century Canal District ringing the Old City to the south and west; and the Jordaan neighborhood to the west (Figure 8.1).

Unlike the surrounding ring of nineteenth-century housing, which comprises nonprofit, social and subsidized housing tenures in various mixes of public and private involvement, the older central and inner city has traditionally been privately owned and owner-occupied. Increasing numbers of vacancies in office, warehouse and even some residential properties in the 1970s could therefore be translated – often via government subsidies – into reinvestment and gentrification. This process was assisted by less strict criteria for functional and tenure conversion after the late 1970s. The Old City proper has certainly experienced this gentrification, but, as in so many other cities, a number of processes and events have combined to limit gentrification in this core of Amsterdam. First, and typically, there is a wide diversity of functions in the center including a concentration of transport, financial, service,

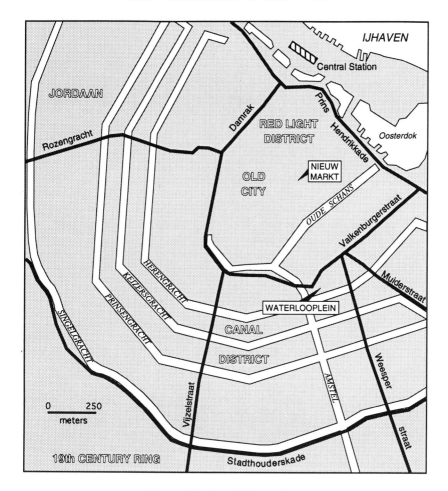

Figure 8.1 Central Amsterdam

government, tourist, commercial, retail and office activities as well as housing; the lower proportion of residential buildings and buoyant rent levels for many non-residential purposes (but not offices, in an international context) limits gentrification at the center. Second, the continued provision of social housing in the Old City – especially in the eastern part of the Old City around Nieuwmarkt (largely rebuilt in the late 1970s and 1980s[2]) and farther east near the port – has also limited possibilities for expanded gentrification. Third, the red-light district dominates a significant proportion of the residential housing of the Old City, and while it is not inherently antithetical to the process, the expansion of prostitution and, in close proximity, street drug trading, has certainly dampened gentrification there.

In fact, as Musterd and van de Ven have argued, the "most evident signs of gentrification are observed in the outer ring of the inner city" (Musterd and van de Ven 1991: 92); that is, in the Canal District and the Jordaan. In this ring, etched by four canals ringing the Old City, private ownership, decentralization and an available if disinvested housing and commercial stock came together to provide the raw materials for the most intense gentrification. Thus in terms of income, "the position of four out of eight inner-city neighborhoods improved relative to the city of Amsterdam" *per se* between 1986 and 1990 (ibid.). This district was originally built in the seventeenth century around four concentric canals as a mixture of merchants' homes and canalside warehouses with retail and commercial outlets often on the ground floor. It is these buildings which were forsaken and devalued as a result of deindustrialization and suburbanization, especially after World War II, and which were available for the beginnings of gentrification in the 1970s and its intensification in the 1980s. Gentrification here "is concentrated along specific streets and canals, which have been totally transformed. . . . Up-market condominiums in restored canal houses or converted warehouses fetch prices of over Dfl. 400,000" [US$275,000] (van Weesep and Wiegersma 1991: 102).

If the process is more accomplished in the Canal District outside the Old City, it is perhaps more symbolic in the Jordaan to the west. The Jordaan was largely built in the seventeenth century, to the west of the Canal District on the fringes of Old Amsterdam. A traditionally middle-class then working-class neighborhood connected to the expanding docks on its northern edge, it experienced long-term disinvestment and physical dilapidation from the late nineteenth century, and by the 1960s, despite geographical proximity, the Jordaan was a socially and culturally isolated enclave *vis-à-vis* the Old City. Cortie *et al.* (1981) found that as late as 1971, 80 percent of properties rented

Plate 8.1 Reinvestment in Amsterdam's Canal District

for less than Dfl. 1,200 (US$220 at prevailing exchange rates) per annum. Average land price in the area was barely 20 percent of that in the adjacent Canal District and city core to the east. By the 1980s, however, despite the smaller apartments of the district, large numbers of professionals, artists, students and other middle-class residents had moved in and storefront premises were increasingly taken over by restaurants, up-market cafes, coffeehouses, art galleries, bookshops and bars. By the late 1980s and early 1990s, the area was totally transformed, so much so that gentrification activity had spilled westward into some of the newer nineteenth-century building stock where private rehabilitation mixes with new construction of social housing.

By comparison with British or US cities, 1980s gentrification in Amsterdam seemed to retain a mix of social classes in the city, at least over the medium term. It was not uncommon to find expensive gentrified buildings adjacent to newly constructed social housing or renovated quarters for the working class; gentrification often occurs "as a piecemeal process concentrated in specific sites." This is especially true of the Jordaan, which retains a strong working-class population amid gentrification. As van Weesep and Wiegersma put it, this "range in prices is quite common in Amsterdam. . . . [T]he rich and the poor live cheek by jowl" (1991: 110, 102). Although the same juxtaposition obviously occurs in any gentrified neighborhood, in the US especially the power of the private market makes this a more transitory situation. In Amsterdam, the medium-term stability of gentrified neighborhoods surely results from the retention of strong state regulation alongside enthusiastic privatization. The vital question surely is: how long can such ecumenical gentrification coexist with traditional working-class habitation of the inner city?

Social opposition to gentrification in Amsterdam, and to housing policies more broadly, has been a staple of the city's politics for the last twenty years. As vacancies increased in and around the central city in the 1970s, young people, whose access to housing was strictly circumscribed amid Amsterdam's housing scarcity and highly regulated housing system, began to squat vacant properties. The Battle of Waterlooplein, in fact, erupted in conjunction with an unprecedented squatters' movement in Amsterdam which received international media coverage. Just as the Amsterdam City Council was initiating its first forays into privatization in early 1980 and beginning its cutbacks in the funding of social housing, it decided also to move the police against a series of squats that had been established in the city. The use of water cannons, tanks as bulldozers, and 500 police in riot gear achieved the temporary eviction of squatters from Vondelstraat outside the southwestern edge of the Canal District in March 1980, but this turned out to be a prelude for a larger confrontation the next month. On April 30, on the occasion of the coronation of Queen Beatrix, a nationally coordinated set of squats, invasions and demonstrations provoked a furious police response, not least in Waterlooplein. Before long the Amsterdam celebrations devolved into a series of running street battles as other anticoronation demonstrators were caught up in the fray and targeted by the police. With tear gas wafting over the city, coronation festivities were rapidly curtailed.

For the next few years, the Amsterdam squatting movement was at its zenith, enjoying considerable if begrudging political clout. It peaked at perhaps 10,000 people, but was clearly losing influence by the end of the 1980s. In October 1992 the "Lucky Luyk" squat, also on the southwest edge of the Old City, was subjected to a surprise Special Squad raid which pre-empted an imminent agreement between squatters and the city council. This again provoked widespread sympathetic support for the squatters and led to three days of the most violent confrontations, which spread into the central city. Barricades, wrecked cars and a tram set alight were met by tear gas and water cannons as the mayor declared a state of emergency that gave police broad powers to arrest anyone on suspicion. The squatters' movement faced a further rebuff when squatters were removed from a central city squat to make way for a new Holiday Inn. Today there are still several thousand squatters in numerous squats around the city but the political power of the movement has dulled in the face of concerted police containment, municipal housing policy and expanding gentrification.

The story of gentrification in Amsterdam, then, is the story of state involvement in the land and housing markets, the relaxation of state controls which precipitated a rush of private investment, and broad, sometimes violent, political opposition to state and private designs over housing. Although the stated aspirations for the 1989 Housing Policy include a more equitable allocation of homes, attendant on the enhanced mobility of higher-income residents (especially those who have till recently lived cheaply in social housing or otherwise low-rent accommodations), even cautious supporters of deregulation and privatization register a certain foreboding about the results. "Current shifts in Dutch housing policy will undoubtedly stimulate gentrification," and although in the short run this may well lead to a greater diversity of social groups in a given neighborhood (van Weesep and Wiegersma 1991: 110),[3] already these policies "may have contributed to the spatial segregation of income groups": the "danger of increased segregation" (van Kempen and van Weesep 1993: 5, 15) in an urban system that previously exhibited comparatively low levels of segregation vis-à-vis others in the advanced capitalist world. The pattern of segregation is familiar: wealthier private residents are more and more concentrated in the central gentrifying districts as well as traditionally better-off neighborhoods to the southwest, while lower-income social housing is increasingly made available on the edge of the city.

In distinction to the US experience where, as in Society Hill (Chapter 6), gentrification was often actively subsidized from the beginning, in Amsterdam the process emerged very much in the interstices of the market and official housing policies. Only later did gentrification become de facto public policy. As Musterd and van de Ven (1991: 92) put it: "the spontaneous process of gentrification of parts of the inner city, which had started in the seventies, became a policy goal. Gentrification was embraced as the lifebuoy for the big city." The policy, they continue, has become almost too successful, leading to the displacement of offices and jobs by luxury apartments.

But it is not only offices but poorer residents who, as in other cities experiencing gentrification, are in danger of displacement. Although active or even passive displacement has been relatively rare, given the strength of tenant protection laws, the rush of gentrification since the 1980s and the relaxing of regulations have raised housing costs dramatically and increased the level of evictions in Amsterdam. While "massive displacement has not yet occurred" in most Dutch cities, van Kempen and van Weesep (1993: 15) suggest in the context of Utrecht that "an increased displacement . . . will result from the growing trend of gentrification."

Gentrification, displacement and segregation together point toward a significantly restructured urban geography. In a script that is very similar to that observed in relation to Harlem (a New York neighborhood which was of course named after a Dutch city), and outlined by Harold Rose (1982; see p. 162 above), Cortie and van Engelsdorp Gastelaars early presented the following scenario:

> If these continue to be the results of future renewal operations then the late nineteenth century zone bordering on the inner city will also become inaccessible for the ethnic minorities, most of whom live in families. This will lead to a concentration of ethnic migrants, and comparable financially weak categories of citizens, on the city fringe and beyond. The result might be that post-war quarters with flats to rent on the outskirts of Amsterdam seem to be pre-destined for social down-grading in the future.
>
> (Cortie and van Engelsdorp Gastelaars 1985: 141)

BUDAPEST: GENTRIFICATION AND THE NEW CAPITALISM

The most dramatic shifts in the Budapest land and housing markets obviously came following the political and economic transformation of 1989, and herein lies the larger theoretical significance of gentrification in Budapest. Insofar as it involves a dramatic, perhaps unprecedented, shift from minimal to maximal investment in a newly evolving land and housing market, it provides a laboratory for examining the interconnected parries of supply and demand, the impetus of production-side and consumption-side forces in the genesis of gentrification. Bearing in mind the very different twentieth-century history of Budapest compared with the cities of North America, Western and Northern Europe, and Australia – economically, politically and culturally – then to the extent that gentrification emerges hand in hand with the capitalization of the Budapest land and housing markets, a radically different housing and institutional history notwithstanding, it may be reasonable to make certain generalizations about gentrification and indeed to treat the process as endemic to a certain stage of late-capitalist urban development.

This argument must be made carefully, however. Hungary's economy, of course, included a considerable private sector even prior to the "liberalization" of the late 1980s. Budapest, for example, with 67 percent of all housing publicly owned in 1975 (Pickvance 1994: 437), actually trailed Dundee and Glasgow in this category. And the privatization of housing was actually incorporated into formal housing policy as early as 1969. The practical effect of these laws was minimal, however, and the state retained strong centralized control over housing construction and allocation, leading to only very low levels of privatization before the early 1980s. At that point, social polarization began to intensify somewhat, leading to an expanded middle class, an expansion of suburban enclaves and price inflation, especially in the Buda Hills. This polarization, together with the class segregatory effects of state rehabilitation and allocation policies dating from the late 1970s, provided an opening for localized instances of what Hegedüs and Tosics (1991) have termed "socialist gentrification."

Whatever the incidence of incipient gentrification *per se* prior to 1989, it should be clear that the prior experience of urban development has been dramatically overwhelmed by rapid and intense reinvestment, especially at the urban core. Budapest in the 1990s is, as Kovács (1994) has put it, a "city at the crossroads." The development of a post-Fordist economic base, already significantly in place in the 1980s, has been accelerated, and the city finds itself the crucible of "a new national modernization process, when new functions (i.e., new forms of capital accumulation) are emerging, and this process is proceeding in a very concentrated way" (Kovács 1994: 1089). Along with the steady privatization of the land and property markets, Budapest has attracted an unprecedented flood of foreign capital, new corporate headquarters and branch offices, joint ventures, corporate service companies, hotels and tourist facilities, and a commercial explosion. Milan clothiers and German banks have shoehorned themselves into the tight central streets or expanded along the boulevards of Budapest alongside the Hilton Hotel and Procter and Gamble. Major European developers have converged on the city (Skanska, Muller AG, Jones Lang Wootton, Universale), and a series of ambitious redevelopment plans, aimed at the inner city, are being drawn up (Kovács 1993). Budapest attracted the lion's share of $5 billion foreign commercial investment in Hungary in the four years after 1989 (Perlez 1993).

In addition, the transforming inner districts of Budapest are attracting the kind of expensive restaurants, clubs and nightspots that mark many gentrifying neighborhoods but which were much rarer in Budapest a decade ago. Whole buildings are being renovated throughout the inner neighborhoods and rare vacant lots are being rebuilt with the offices that are increasingly integrating the city and the Hungarian economy into global capitalism. Although a ban on large skyscrapers has halted some projects – most notoriously perhaps a forty-storey office block planned by French developers – escalating investment has created an unprecedented and highly concentrated demand for land and space. By 1993, monthly office rents in the central city had increased to between $30 and $38 per square meter – not yet New York or London levels, but rivaling those in Vienna and double the rent levels in Amsterdam and Brussels (Kovács 1993: 8).

Gentrification is integral to this changing social, economic and political geography of Budapest occurring at the behest of global integration. Unlike that in London or New York prior to the 1970s, then, Budapest gentrification did not begin as a largely isolated process in the housing market, but came fully fledged in the arteries of global capital following 1989.

Plate 8.2 Office and residential reconstruction on Nyár ut, in District VII, Budapest
(Violetta Zentai)

The opportunity for gentrification in Budapest lies in a protracted history of disinvestment. But in this case, the disinvestment has a quite different origin from the disinvestment that served to put Society Hill or Harlem or Old Amsterdam on the gentrification platter. Public policy was the key here. In direct contrast to what happened in the US and Western Europe, where, even in the case of Amsterdam, the heaviest state regulation never supplanted private housing construction, in Budapest the postwar suburban expansion was state led and was dominated by social housing. This is not to say that the private market was nonexistent; rather it is to suggest that the considerable private suburbanization that did occur took place more in the interstices of state regulation and policies than the other way round. Postwar housing policy was focused primarily upon resolving a major housing crisis for the poorest part of the population, and this meant the construction of a large number of small and simple social housing units on the outskirts of the existing city. "A disproportionately large part of public expenditure was spent at the edge of the city throughout the 1960s and 1970s," Kovács (1994: 1086) reports, and consequently "the entire central-city was neglected for a long time." If the causes were more state-centered than market-centered, the spatial result – at least in terms of the geography of investment and disinvestment – closely paralleled the result in the US especially: investment at the outskirts, disinvestment at the center. By 1989, most of the urban core had experienced no significant reinvestment for at least half a century and it was not uncommon, even in the more fashionable neighborhoods, to see buildings pockmarked with bullet holes, unrepaired since the 1956 revolt.

If overall state regulation of the land and property markets was already being relaxed in the 1980s, constraints on private land and housing trans- actions were largely removed by Parliament after 1989, bringing about a rapid privatization. An already chronic shortage of housing and high-density overcrowding were exacerbated. By the summer of 1993 approximately 35 percent of the public housing stock had been privatized, and the public sector now accounted for only 32 percent of all housing (Hegedüs and Tosics 1993; Pickvance 1994). This privatization was concentrated at the top end of the market (Kovács 1993), first and most powerfully in Buda, but by the early 1990s privatization also dominated property market transactions in inner Pest. It has led to a quite evident class shift in the composition of several Budapest neighborhoods which are attracting large numbers of the newly emerging middle class. This gentrification is focused in several central districts, but as with gentrification in most cities it is highly visible, involves dispropor- tionate amounts of new capital investment, and already has the momentum to become a major determinant of the new urban landscape.

The most evident transformation is taking place in an arc around the government and business centers on the Pest side of the Danube (Figure 8.2). In the inner neighborhoods of Erzsébetváros (District VII), and Terézváros (the old Jewish ghetto – District VI), as well as the central District V in Pest, and to a lesser extent in the Old City of Buda, there has been an explosive

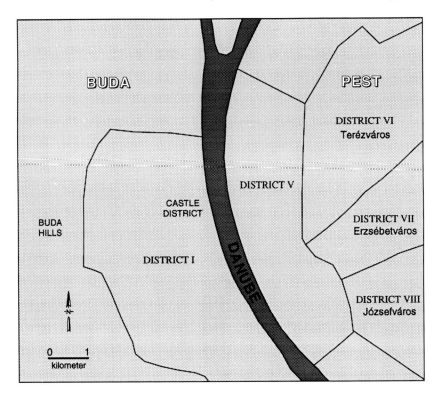

Figure 8.2 Gentrifying districts of Budapest

level of reinvestment leading out from the traditional city center. In these neighborhoods especially, the effects of disinvestment resulting from decades of nonmaintenance and minimal reinvestment were exacerbated by Hungary's full insertion into the global market, and the sudden confrontation with land and housing prices in other global centers. Previously state-owned apartments and apartment buildings have been privatized; and newly privatized apartments and buildings have been renovated, this in many cases being the first significant reinvestment since World War II. Other privatized apartments have been sold and resold as the sudden introduction of a comparatively free market has led to rapidly rising land and housing prices. Between 1989 and 1991, prices within the districts of this arc had risen by between 52.1 percent and 81.1 percent, clearly outstripping every other district in the city except for the Buda Hills, which continued to experience comparable price increases. Since 1991 the price inflation has been even more striking; with prices at unprecedented levels by the mid-1990s, those tenants who had the resources to buy in the first place or who had the greater resources to get in later could do so with the anticipation of considerable (speculative) financial gain.

Whatever the obvious connections between gentrification and the corporate and commercial expansion of the global market into the inner neighborhoods, the wholesale gentrification of Budapest – for all of its dramatic momentum amassed in barely a decade – should not be taken entirely for granted. Various economic, social and political forces might well limit the process. These limits revolve around issues of demand, supply of mortgage capital, and existing patterns of ownership, and they will be considered in turn.

In the first place, there is the question of demand. In Western Europe, North America, and Australia, neoclassical economists notwithstanding, demand is relatively unimportant as a limiting factor for gentrification. As we saw in the case of Harlem, it was possible to infuse demand into a depressed market if the properties could be put on the market; at least in comparison with Hungary, the demand is spontaneously there, and becomes an issue only at the temporal or geographical margins of capital accumulation, for example in smaller cities or during severe economic depressions. In Hungary, however, that level of "conventional demand" is only now being organized as a social given, and it is not possible to assume quite such an automatic market for gentrified apartments, blocks and neighborhoods. Where privatization and escalating investment have opened up a potential reservoir of new residences, the continuation of gentrification is also dependent on the equally rapid emergence of a more differentiated class structure than existed prior to 1989, and in particular the expansion of a middle and upper middle class whose incomes make them the potential source of gentrifiers. That kind of class differentiation is indeed occurring as the rapid expansion of corporate offices and retail businesses around the traditional core spawns a comparatively wealthy professional and managerial class earning salaries pegged to the global market. And yet insofar as it accomplishes the conversion of inner-district buildings to offices, this same corporate expansion will simultaneously absorb residential buildings for office functions, thereby diminishing the supply of properties available for residential gentrification. The geographical results of all this remain to be seen.

The invasion of corporate office development into residential neighborhoods might restrict gentrification, but there is no reason to assume that it would not, as in many other cities, simply displace the gentrification frontier outward. In either case the question of demand is for the moment tightly bound up with the rhythms of global investment in Budapest.

A second, related, and potentially more important institutional constraint may limit the economic and geographical expansion of gentrification. It is a constraint that more broadly affects the privatization of housing. Despite the significant private market in housing prior to 1989, no significant lending institutions have developed for the provision of mortgage credit. Mortgage credit can be thought of as the glue provided by the suppliers of capital to ensure that demand keeps up with supply. By the mid-1990s, with no significant mortgage credit in place, the consumption side of gentrification was funded in one of three main ways. Some individuals had saved sufficient funds to rehabilitate their own apartments, or to buy or rent previously refurbished quarters. Alternatively, personal funds could be augmented with loans from regular money lenders at interest rates of 30–35 percent in the mid-1990s. Or third, the construction and rehabilitation, rental or purchase of luxury accommodations in central Budapest might be underwritten by national and international corporations expressly for their employees. In addition, gentrified apartments might simply be sold (or more likely rented) to foreign professionals whose work brings them to Budapest and who have the requisite funds or income. Whatever the lack of mortgage financing, other sources of capital have so far proven plentiful enough to facilitate the explosion of gentrification in central Budapest in the early 1990s. Between the private market and Parliament there are energetic attempts to generate mortgage lending institutions which would significantly lubricate gentrification.

Finally, existing patterns of ownership and control might limit the extent and nature of gentrification. On the one hand, the privatization of social housing since the late 1980s means that as regards any specific apartment building, a would-be developer often has to deal with multiple apartment-owners, thus often limiting the feasibility of renovations; apartment owners may not have the wherewithal to renovate yet may be reluctant to sell out in an upwardly spiralling market. On the other hand, there are pockets of continued social ownership in the Budapest land market, especially in the Castle district which, with its old stone houses, has already attracted some gentrification. The complete gentrification of this area will presumably depend on the privatization of the housing stock.

These constraints are real enough, but they have not decisively hampered a gentrification of inner Budapest that has been nothing short of explosive. "Most of the essential preconditions of Western-type gentrification apply equally well in the case of Budapest" (Kovács 1994: 1096). Whereas prior to the early 1980s, state regulation of land and housing prices prevented a significant rent gap from materializing despite wholesale disinvestment – capitalized and potential ground rents were strictly regulated and kept low – the recapitalization of the Budapest market since the late 1980s has both mobilized existing disinvestment *vis-à-vis* the new market and inflated prices, and therefore the potential ground rent, to produce a significant rent gap (Hegedüs and Tosics 1991: 135). Indeed, not just in Budapest but in the secondary cities of Hungary, a similar conclusion can be reached. In their research on Békéscsaba, Beluszky and Timár could already report in 1992 that:

> Privatisation and the growth of a rent gap and value gap have created the possibilities for a Western-type gentrification in some new flats built on the site of old inner-city residential areas. . . . In 1991 in a downtown block in Békéscsaba with 30 flats, half the buyers, a proportion never seen before, belonged to the intelligentsia, entrepreneurs and white-collar workers.
>
> (Beluszky and Timár 1992: 388)

The future of gentrification will also be framed by the state's housing policy. Entire areas of Budapest, most notably the Old City of Buda, which would be an ideal target for gentrification, remain largely in public ownership, and the possibilities for gentrification there, while not entirely determined by public policy, are largely dependent upon it. Although the housing sector is now open to privatization, different districts in the city have moved forward with privatization programs at very different rates. The momentum behind housing privatization and the extent to which the state remains committed to providing affordable housing for the working class will be crucial. Beluszky and Timár (1992: 388) suggest that if the broad social rationale of current housing policy is not dismantled, then there is "the possibility of avoiding a New York–type . . . gentrification" dominated by private investment. They suggest instead that the more piecemeal geography of gentrification as found in Amsterdam may be a more appropriate model, and indeed in the first few years of major gentrification in Budapest and elsewhere in Hungary, the process is dominated by piecemeal and infill redevelopment. Less optimistic is Kovács (1993: 8):

> The best example of high-spired privatisation is the very heart of the CBD, the 5th district. Although local government leaders repeatedly stressed that the district [does] not intend to get rid of its social housing, yet there are signs that the process cannot be stopped and the whole district will be privatised in a short time. In 1989, approximately 55% of the district's social dwellings were placed on a prohibition list, which meant that these flats should have been preserved in state (i.e. local government) ownership. However, this list was re-examined and revised from time to time, and by the middle of 1992, already 65% of the flats were virtually sold.

As elsewhere, gentrification in Budapest will have a deleterious effect on those who are unable to afford the refashioned apartment blocks and neighborhoods. "There is no doubt" that many remaining local residents "will have to leave the centre after a while, since escalating rents and property values will edge out the poorer families" (Kovács 1994: 1096). Indeed displacement has already begun. Homelessness, which was virtually unknown in Hungary, suddenly became an issue as small numbers of homeless people began congregating in the Budapest train station in the late autumn of 1989. In large part they had been forced out of housing following the first round of privatization and decontrol measures passed by the Hungarian Parliament in the late summer of that year. By 1994, official estimates of homelessness nationwide were approximately 20,000, but unofficially there was wide agreement that the figure was much higher. A government official speculated that the figure might even be 120,000, but that there was little way of knowing. Rising unemployment (estimated at over 13 percent in 1994) clearly also contributed to burgeoning homelessness. Elsewhere, most notably in District VIII where much less housing has been released from the public sector, ghettoization rather than gentrification is the dominant process.

And much more significant displacement is likely. To take just one example, the $1.2 billion Medach-setany "Business District" development in District VI is planned to include offices, shops, a mall and luxury apartments. It will demolish a swath of slum buildings, opening up a wide pedestrian street through the neighborhood (Kovács 1993: 9). While the mayor of the district supports the project, arguing that it will "enhance property values" and "revalue the district," others have opposed it. Estimates range from 522 apartments demolished, displacing over a thousand people, to three times those figures (Perlez 1993).

This of course is akin to the scale and style of Western redevelopment projects, whether at Society Hill Towers or Amsterdam's Waterlooplein. What distinguishes Budapest in the 1990s, however, is that despite rampant reinvestment in poor and working-class neighborhoods, evictions, unprecedented price and rental levels, and a rising homeless population, there has been very little resistance or organized opposition to gentrification. Significant resistance emerged in the late 1970s surrounding several state-sponsored redevelopment projects, but at a public hearing in the early 1990s to introduce the massive Medach-setany "Business District" development, an overflowing local audience was certainly vociferous but organized opposition minimal. Whatever the broad discontent concerning the housing crisis in the "period of transition," political futures beyond those delineated by the market are radically foreclosed in the official political discourse, and by the mid-1990s opposition to market-induced social effects in general has not coalesced. Nothing of the experience of Amsterdam or, as we shall see, Paris or New York has in that regard yet been seen in Budapest. The major political battle, so far, has centered on the extent to which housing and other social regulation ought to be relaxed.

PARIS: DEFERRED AND DECENTRALIZED GENTRIFICATION

If in some cities – Budapest and Amsterdam are good examples – gentrification tends to be quite centralized around the urban core, in Paris it has affected a variety of neighborhoods scattered around the city. Yet like Amsterdam, much Parisian gentrification affects neighborhoods in which the building stock dates back several centuries and which have experienced much longer bouts of disinvestment (and at times reinvestment) than US cities or, for that matter, some European cities such as London. The latter has a long urban history to be sure, but most of the inner-London neighborhoods experiencing gentrification owe the majority of their building stock to the late eighteenh or more likely the nineteenth century. In several Paris neighborhoods, the stock currently being gentrified was originally constructed in the late sixteenth through the early eighteenth centuries as tenements. In respect of its housing stock, then, Paris has more in common with Edinburgh than London. This remaining stock has survived various forms of creative destruction (nineteenth-century "embourgeoisement" of the city) and not so creative destruction (wars).

In each of the neighborhoods where gentrification has emerged, it has taken a particular form. Perhaps the most classical gentrification has indeed taken place closest to the center. On the Île de la Cité and Île Saint-Louis, for example, in the middle of the River Seine, old sixteenth- and seventeenth-century apartment buildings that had fallen into gradual disrepair since before the Paris Commune of 1870–1871 began in the 1960s to experience sustained reinvestment; so much so that an Île address today is one of the most sought-after and most expensive in Paris (Figure 8.3).

On the Left Bank, it is mass tourism on top of an existing intellectual and artistic culture that has dramatically propelled the old Latin Quarter, also as much as four centuries old, into the circuits of international capital. The once dingy streets, dotted by smoky cafes, bookshops and traditional restaurants,

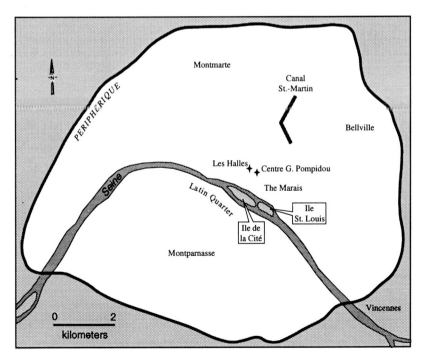

Figure 8.3 Paris gentrification

are now splashed with brightly colored boutiques and postcard vendors, delicacy food stores and upscale cafes, a global bazaar of fast foods and tourist-oriented "French" and Greek restaurants. If the streets of the Latin Quarter, especially around Saint-Michel, are now thronged with non-Parisians, the apartments in the densely built fifth, sixth and seventh *arrondissements* are now occupied largely by professionals, artists and students who are or have become Parisians, as well as older residents. A small two-room garret which in the early 1970s was on the market for FF 20,000 or FF 30,000 might sell in the mid-1990s for ten or fifteen times that amount, while larger apartments neighboring the boulevard Saint-Germain could easily fetch FF3–5 million (US$800,000–1m). These properties are as likely to be advertised by Sotheby's or Century 21 International as by the more elite brokers of Paris.

Equally gentrified though somewhat more removed from the tourist mainstream is the Marais on the Right Bank. Since the nineteenth century a working-class neighborhood with something of a bohemian reputation, an old Jewish quarter, and more recently the home to numerous immigrants, the Marais had been a vibrant and fashionable district housing members of the aristocracy and an emerging bourgeoisie prior to the 1789 Revolution. Widely barricaded in the Revolution, it was largely abandoned to the Parisian working class and to a later Jewish population, and the housing stock of four- or five-storey buildings increasingly subdivided. It has therefore undergone protracted if not catastrophic disinvestment for almost two centuries. The construction in the 1970s of the Pompidou Center and the replacement of the old Les Halles meat market by a modern complex of cultural and consumption venues on

Plate 8.3 A street in the Marais, Paris

the western edge of the Marais galvanized a significant reinvestment in the Marais, which now hosts "an eclectic mix of the old and the new, as high-class fashion outlets have taken over former food shops, often retain[ing] the former 'boulangerie' or 'patisserie' signs" (Carpenter and Lees 1995: 298). The renovated housing stock now attracts a wide range of young and not so young urban professionals, but its gentrification has been both less complete and less boisterous than that in the Latin Quarter. The Jewish section of the Marais especially retains some of its traditional appearance and ways in the face of the financial pressures of gentrification.

Montmartre also experienced significant disinvestment, at least after the Commune of 1870–1871, although this was significantly circumscribed by the continued symbolic political importance of this district of the city. Yet here too, tourism, relative altitude and a certain artistic nostalgia, all inscribed into Montmartre's landscape, have encouraged gentrification. Elsewhere – around Montparnasse on the Left Bank, for example, or on the Right Bank along the southern stretches of the Canal St. Martin – the newer nineteenth-century building stock has also seen patchy gentrification. In the former case, Montmartre was always a bohemian neighborhood attracting significant numbers of intellectuals and artists while the gentrification of the Canal St. Martin owes more to the revaluation of canal-side property following deindustrialization and the folding of the many small craft shops and businesses that clustered round the canal. Even in Belleville, a solidly working-class neighborhood in the northeastern outskirts of Paris, now a major center of Arab, African, Chinese and East European immigration and a traditional stronghold of proletarian opposition, Parisian city authorities have initiated a series of "amelioration schemes" aimed at gentrification.

With reference to Claval (1981), Carpenter and Lees (1995: 292) make the point that with some exceptions (such as the Marais), central Paris "has always been dominated by the upper classes" and that when one combines this with the fact that its suburbs are predominantly working and lower middle class, Paris presents "the direct inverse of the traditional Anglo-Saxon concentric-zone social structure model developed by the Chicago School." While this may be a rather hasty dichotomy between an Anglo-Saxon urbanism and the rest – the concentric ring model barely applied to New York and even less to London, even in the industrial era it depicts, as they too retained significant affluent areas at the center and large working-class suburbs – it nevertheless points to an important feature of the historical geography of Paris. In fact the Revolutions of 1789 and 1848 together with the Paris Commune of 1870–1871 all convinced the French ruling classes of the need to reclaim the central city and disperse the working classes to the urban edge (Pinkney 1957; Harvey 1985a). This helps to explain the rela-tively dispersed geography of gentrifying neighborhoods in Paris today. Many but not all of the affected areas represent districts that escaped Baron Haussmann's onslaught after 1851, the destruction following the Commune, and like efforts.

For all of the intense class geography of central Paris, it is not an accident that the term "gentrification" was coined across the Channel. Rapid and highly visible gentrification only really came to Paris at the beginning of the 1980s, somewhat lagging most comparable West European cities. The issue was not so much levels of disinvestment as of reinvestment. The central connection between gentrification and disinvestment and reinvestment is well understood in the French context: "The phenomena of the deterioration and improvement of old districts are the social processes concretized in the evolutionary movement of devaluation and revaluation" (Vervaeke and Lefebvre 1986: 17; my translation). Rather, the deferral of gentrification resulted from a conjuncture of financial, institutional and tenurial circumstances.

First, in an ironic parallel with Budapest today, postwar Paris did not have a well-developed mortgage infrastructure. This was largely due to tight state restrictions on France's powerful and highly centralized banking capital. But also, and related, Paris has traditionally had a large private rental sector with barely a quarter of its housing units in owner-occupancy. With its ancestral power invested in *rentier* capital, the *petit bourgeoisie* feared the erosion (as in London, for example) of the city's traditional apartment rental market, and, despite the hard-fought passage of several housing acts aimed at progressively easing state restrictions, the market did not open up until the late 1970s. Third, almost a quarter of the Paris housing stock was rent controlled as late as 1978 and this also contributed to "postponing gentrification" (Carpenter and Lees 1995: 294). Prior to the late 1970s, cautious deregulation had almost exclusively benefited suburban development, but a 1977 Housing Act offering low-interest mortgage capital to prospective home and apartment owners significantly fueled the gentrification of several Paris neighborhoods.

Opposition to gentrification in Paris, as in Amsterdam or New York's Lower East Side, has been loosely integrated as part of a larger housing and anti-homelessness movement. The election of the Gaullist government in 1986 led to the enthusiastic ideological advocacy of private-market policies which had been begrudgingly implemented by the previous socialist administration, and have been continued by successive socialist and Gaullist governments alike. Prices, speculation and numbers of evictions all soared, in a more dramatic inflation of the Paris housing market than affected even New York. This provoked an immediate response in the streets. Since the late 1980s, thousands of protestors have set up various encampments in different parts of Paris. They have demanded that an estimated 65,000 Parisians who are badly housed, and the city's homeless people (numbering at least 20,000), be given access to vacant apartments, which have been estimated to number as many as 117,000 ("Les sans-abri . . . " 1991). There have been loosely linked protests against a series of publicly sponsored developments, going back to Les Halles in the early 1970s, which led to housing clearance and gentrification. The situation intensified through the 1980s when gentrification spurred widespread speculation, and rental levels increased by a factor of between five and ten in some districts; landlords took to evicting tenants, often immigrants, in search of higher rents.

One of the most publicized and most determined protests began in July 1991 when thirty-seven homeless families constructed a shanty town on the declared site of the new Grande Bibliothèque near the Gare d'Austerlitz

(Nundy 1991). Largely led by immigrant African women, dwellers in the tent city refused to disperse as their numbers grew to over 400. They demanded decent housing, citing the 110,000 apartments which, according to official figures, lie vacant in Paris as landlords speculate on rising rents. They were eventually forcibly removed from the site. A later shantytown in Vincennes, on the city's eastern outskirts, housed 300 families. Like the first group, the Vincennes squatters were offered inferior housing in various prefectures outside Paris; just as quickly as the protestors refused the accommodations, various local mayors vowed to prevent their moving in ("Les sans-logis . . . " 1992; "Les Africains . . . " 1992). At a later incarnation of these encampments, several hundred homeless people in a tent city outside the Social Affairs Ministry, were evicted by riot police in a "surprise" dawn raid on 13 December 1993. The raid followed the deaths of ten "destitute" people in Paris "during a cold spell" in November. Homeless people were "chased, treated brutally, handcuffed and insulted" by Paris police, according to a statement in the *New York Times* ("After eviction . . . " 1993).

CONCLUSION

I have argued elsewhere that the dichotomy between a European and North American experience of gentrification may be a false one (N. Smith 1991), and I think these and the prior case studies confirm the conclusion that there is as much differentiation of the gentrification experience within Europe or North America as between them and that a continental "divide" would not be useful (see also Carpenter and Lees 1995). The importance of the state and squatter opposition in Amsterdam; gentrification as an integral part of Budapest's new capitalism; the deferred and geographically decentralized gentrification of Paris – these are quite different stories of gentrification. This is not to suggest that there are no coherent transatlantic contours to the different experiences of gentrification; some such differences do exist. The gentrification process in the US especially does represent a certain extreme vis-à-vis Europe; it is typically faster, more widespread, more complete in affected neighborhoods; it is premised on much more severe attacks of disinvestment, and may lead to more dramatic shifts in investment patterns and urban cultures. And there are clearly political reasons why one might want to emphasize a certain US exceptionalism. Yet these general differences really do not gel into a sustainable thesis that these are radically different experiences. It is difficult to argue that New York, with a vast landscape of gentrification, has less in common with Amsterdam than with Houston or Los Angeles; that downtown disinvestment in Phoenix is more severe than in Glasgow or Lille; that the new capitalism in Budapest more resembles that in Paris or the cautious gentrification of Oslo than it does that of Philadelphia; or that anywhere in the US has seen a faster gentrification than Budapest since 1989. In terms of gentrification, I suspect that Paris and Amsterdam, for all their own differences, have more in common with Baltimore and Seattle than with Rotterdam and Rome.

By the same token the existence of difference is a different matter from the denial of plausible generalization. I do not think that it makes sense to dissolve all of these experiences into radically different empirical phenomena. It seems to me that it is of primary importance to retain a certain scalar tension between, on the one hand, the individuality of gentrification in specific cities, neighborhoods, even blocks, and on the other hand a general set of conditions and causes (not every one of which may always and necessarily be present) which have led to the appearance of gentrification across several continents, at approximately the same time. The power of a more general theoretical stance is augmented by the suppleness that comes from a sensitivity to the details of local experience – and vice versa.

This was driven home to me on a recent Saturday afternoon as I sat in a gentrified bar in the Marais. It was an Australian bar and I was drinking Dutch beer, paid for in French francs of course. The few bar patrons, if not French, were English Parisians or German or Japanese tourists. On the television was American football (European League) with Barcelona beating Scotland – heavily. Actually, all of the players seemed to be from the US. The two commentators were Irish and American. It all seemed very familiar, this gentrification of the Marais.

Part III

THE REVANCHIST CITY

MAPPING THE GENTRIFICATION FRONTIER

The main point is that you want to be out on the frontier of gentrification. So you can't use the established financial institutions, for example, the banks. That's why you need the broker. . . . You try to be far enough out on the "line" that you can make a killing; not too far where you can't offload the building but far enough that you can buy the building cheap enough to still make money and not get scalped.

(Steve Bass, Brooklyn developer, 1986)

The irresistible appeal in the press and the public to script gentrification as a new frontier comes from many sources. It is a highly resonant imagery bound up with economic progress and historical destiny, rugged individualism and the romance of danger, national optimism, race and class superiority. But it also comes from the geographical specificity of the frontier. The frontier of the American West was a real place; you could go there and virtually see the line, as Frederick Jackson Turner put it, between "savagery and civilization." The geography of the frontier was cast and created as a container of all these accumulated meanings; the sharpness of the geographical frontier was an excellent conveyance for the social differences between "us" and "them," the historical difference between past and future, the economic difference between existing market and profitable opportunity. This dense layering of meanings is expressed sharply in the shifting frontier line itself.

Much the same is true of the new urban frontier. Whatever the cultural verve and optimism with which the city is seen as frontier, the imagery works precisely because it manages to express all of these meanings in a single place. That place is the gentrification frontier. The gentrification frontier absorbs and retransmits the distilled optimism of a new city, the promise of economic opportunity, the twin thrills of romance and rapacity; it is the place where the future will be made. This cultural resonance comes to make the place but the place is made available as a frontier by the existence of a very sharp economic line in the landscape. Behind the line, civilization and profit-making are taking their toll; in front of the line, savagery, promise and opportunity still stalk the landscape.

This "frontier of profitability," invested with such a wealth of cultural expectation, is a viscerally real place inscribed in the urban landscape of gentrified neighborhoods. In fact it can be mapped. And by mapping it I hope to expose the ideology of "the new urban frontier" which has been so effectively mobilized to justify not just gentrification but urban restructuring more broadly. For behind the heavily laden and redolent cultural appeals to a "new urban frontier" lies a more prosaic economic truth that gives the frontier imagery the semblance of legitimacy.

BENEFITS OF DISINVESTMENT

As an economic line, the gentrification frontier is sharply perceived in the minds of developers active in a neighborhood. From one block to the next, developers find themselves in very different economic worlds with very different prospects. The "gentrification frontier" actually represents a line dividing areas of disinvestment from areas of reinvestment in the urban landscape. Disinvestment involves the absolute or relative withdrawal of capital from the built environment, and can take many forms. Reinvestment involves the return of capital to landscapes and structures that previously experienced disinvestment. Ahead of the frontier line, properties are still experiencing disinvestment and devalorization, through the withdrawal of capital or physical destruction, owner-occupiers, financial institutions, tenants and the state. Behind the frontier line, some forms of reinvestment have begun to supplant disinvestment. The forms taken by reinvestment can vary substantially; it may involve private rehabilitation of the housing stock or public reinvestment in infrastructure, corporate or other private investment in new construction or merely speculative investment involving little or no physical alteration of the built landscape. Thus conceived, the frontier line represents the leading historical and geographical edge of urban restructuring and gentrification.

Mapping the frontier line – establishing its shifting location – not only provides a means for mapping the spread of gentrification, but also provides a tool through which local neighborhood organizations, residents and housing activists can anticipate gentrification and thereby defend themselves against the processes and activities that convert their communities into a new urban frontier. I shall take a detailed look at the formation and spread of the gentrification frontier line in New York's Lower East Side, especially the northern section, dubbed the East Village, where residential rehabilitation of the classic gentrification sort has been occurring since the 1970s. This neighborhood was deliberately chosen for its inestimable social and cultural diversity and for its intense political opposition to gentrification. To the extent that one can detect an economic geographical regularity to gentrification in the Lower East Side, mapping the frontier will highlight the economic processes that render the cultural ideology of the frontier plausible.

Disinvestment in urban real estate develops a certain momentum that gives the appearance of being self-fulfilling. Historical decline in a neighborhood's real estate market provokes further decline since the ground rent that can be appropriated at a given site depends not only upon the level of investment on the site itself but on the physical and economic condition of surrounding structures and wider local investment trends. It is irrational for any real estate investor – from the owner of a single home to the multinational developer – to commit large amounts of capital to the maintenance of a pristine building

Plate 9.1 Contesting the frontier: abandoned and squatted buildings on Thirteenth Street, Lower East Side, New York (Lauri Mannermaa)

stock amid neighborhood deterioration and devalorization. The opposite process, sustained neighborhood reinvestment, can appear equally self-fulfilling, for it is equally irrational for a housing entrepreneur to maintain a building in dilapidated condition amid widespread neighborhood rehabilitation and recapitalization. In the first case, investment in an isolated building may indeed raise its intrinsic value but it does little to enhance the ground rent at the site and is in all probability unrecoverable insofar as the neighborhood ground rent or resale levels will not sustain the necessary rise in the rent or price of the individual refurbished building. What benefits do accrue beyond the building itself are dissipated throughout a declining neighborhood. In the second case, the neighborhood-wide increases in ground rent that accrue from widespread recapitalization are only partially realized by the owners of buildings who do not rehabilitate their property, although of course short-term speculative gains can be made by warehousing (keeping a building off the market while its price rises), flipping (buying a building in order simply to resell at a higher price) and other speculative practices that involve no significant reinvestment.

Yet whatever the self-contained economic momentum established in neighborhood decline, that decline does not result from some irrational psychology endemic to real estate investors. Rather, sustained disinvestment begins as a result of largely rational decisions by owners, landlords, local and national governments and an array of financial institutions (Bradford and Rubinowitz 1975). These represent the major groups of capital investors in the built landscape and they experience various levels of choice in their investment strategies and decisions, and different rationales for investment, most obviously between private-market and state institutions and individuals. For any participant in the real estate market, the level of investment, type of building(s), age of structure, and geographical and market location are all contingent – much more so for financial institutions and landlords than for the homeowner, whose economic investment is simultaneously the physical commitment to a home. Whether to relinquish real estate entirely in favor of other investments – stocks and bonds, money markets, foreign currency, stock and commodity futures, precious metals – is equally an option dependent upon expected rates of profit or interest. The economic effects of state policy are also differentiated according to building and neighborhood characteristics as well as location. However disparate these individual decisions may be (Galster 1987), they represent a broadly rational if not always parallel or predictable set of responses to existing neighborhood conditions.

Whatever the dysfunctional social consequences provoked or exacerbated by disinvestment – deteriorating housing conditions, increased hazards to residents' health, community destruction, the ghettoization of crime, loss of housing stock, increased homelessness – disinvestment is also economically functional within the housing market and can be conceived as an integral dimension of the uneven development of urban place (see Chapter 3). Focusing on the relationship between housing demand and state policy, Anthony Downs (1982: 35) makes this general point when he observes that "a certain amount of neighborhood deterioration is an essential part of urban development." In addition to the effects of state policies, others have highlighted the role of financial institutions in disinvestment and redlining (Harvey and Chaterjee 1974; Boddy 1980; Bartelt 1979; Wolfe et al. 1980) and eventually abandonment (Sternlieb and Burchell 1973: xvi). The ultimate rationale for geographically selective disinvestment on the part of banks, savings and loans organizations and other financial institutions is to restrict the effects of devalorization, economic decline and asset loss to clearly circumscribed neighborhoods and thereby protect the integrity of mortgage loans in other areas. Attempts to delimit the geography of disinvestment serve to circumscribe the social and geographical extent of its economic impact.

Disinvestment has a certain functionality for landlords as well as for financial institutions. While Sternlieb and Burchell (1973: xvi) have argued that landlords in declining neighborhoods, squeezed between decreasing rent rolls and increasing costs, are as much the victims of disinvestment as its perpetrators, there is also a more voluntary dimension to many landlords' involvement in disinvesting properties and devaluing neighborhoods. According to Pater Salins, "most of the present and future owners of this kind of property are there by choice, and are making money." Market rationality together with state policies have "led housing entrepreneurs to make money in ways that involve the destruction of the housing stock" (Salins 1981: 5–6). In the context of New York City, Salins notes that in the process of "graduated disinvestment', building owners become "increasingly exploitative of the property." The building is "milked" of its rent rolls while the landlord progressively reduces and may even terminate the payment of debt service, insurance and property taxes, the performance of maintenance and repairs, and the provision of vital services such as water, heat and elevators. In all likelihood a building at this stage of decline also changes hands frequently, often without the benefit of formal mortgage sources, before ending up in the hands of a landlord who specializes in "finishing." The "finisher" performs the final gutting of the building's economic value up to and including the removal of fixtures, copper boilers and piping, and furniture, which are scavenged for use elsewhere or for sale (Stevenson 1980: 79). Although some landlords who specialize in this stage of disinvestment hire "enforcers" to collect rent from understandably reluctant tenants, others may be more interested in scavenging the building over a short period than in collecting rents. Physical and economic abandonment and landlord-instigated arson-for-insurance have been the eventual fate of many buildings, and New York City went through a spate of such disinvestment from the late 1960s to the late 1970s, synchronized both with the New York City fiscal crisis and with a series of wider, national and global economic recessions and depressions.

Lake (1979) corroborates this view with an intensive empirical examination of landlord tax delinquency in Pittsburgh. He points out a variety of disinvestment strategies, the deployment of which depends on the type of owner (landlord or homeowner), the size of an owner's holdings, the investor's perception of neighborhood property values, and so forth. Lake identifies a "cycle of delinquency" whereby property maintenance, property value and vacancy rates spiral downward in close relationship to each other.

At a more aggregate level, it is possible to view disinvestment as a necessary if not sufficient condition for the onset of gentrification. As argued in chapter 3, it is this sustained disinvestment by landlords and financial institutions that results in the emergence of a "rent gap" between, on the one side, the currently capitalized ground rent under present use and, on the other, the potential ground rent that could be appropriated with the conversion of the neighborhood building stock to a higher and better use through reinvestment in gentrification (Clark 1987; Badcock 1989).

No matter how trenchant or apparently self-fulfilling, therefore, neighborhood disinvestment is reversible. There is nothing natural or inevitable about disinvestment. The fallacy in the "self-fulfilling" thesis arises with the fallacious "assumption that the ownership sector's expectations do not represent an economically accurate response to the dynamics . . . of housing destruction" (Salins 1981: 7). Just as disinvestment and reinvestment are active processes carried out by more or less rational investors in response to existing conditions and changes in the housing market, the reversal of disinvestment is equally deliberate. The individual decision by an investor or housing developer to reverse direction and to embark on a course of reinvestment rather than disinvestment may result from myriad kinds of information and perceptions, from the data of the Real Estate Board or the words of an astrologer. But assuming individual investors do not control the housing market in entire neighborhoods, successful reinvestment is also contingent upon the broadly parallel actions of a range of individual investors. Whatever the individual perceptions and predilections of a given landlord, developer or financial lender, a successful neighborhood reinvestment reflects a rational *collective* assessment of the profitable opportunities created by disinvestment and the emergence of the rent gap. The more knowledgeable, the more perceptive or simply the luckier investor may make the largest returns by responding more quickly, more accurately or even more imaginatively to the opportunities represented by the rent gap, while the less knowledgeable, the less lucky and the inappropriately imaginative investor may misjudge the opportunity, making lower profits or even sustaining a loss. But in reversing the market logic of decline, all of these actors are responding to a situation already established by the structured actions of myriad investors over a number of years, even decades.

WINDFALLS OF DELINQUENCY: TAX ARREARS AND TURNING POINTS

In the history of neighborhoods, there are not always sharp reversal points or turning points from periods of investment to disinvestment or vice versa. Many neighborhoods receive a more or less steady supply of necessary funds for financing repairs, maintenance and building transfers, and therefore do not experience sustained disinvestment and physical decline. What interests us here, however, are those cases where steady reinvestment has not occurred, where a neighborhood's building stock is thereby devalued, and where gentrification-related reinvestment is initiated – those moments when relatively sharp changes do actually occur in the pattern of investment and disinvestment in a neighborhood. This therefore represents a very specific aspect of the economic history of certain neighborhoods – what we might refer to as the "turning point" where disinvestment is succeeded by reinvestment. Dating this turning point when reinvestment supplants disinvestment in a neighborhood would give a quite sharp temporal indicator of the onset of gentrification-related activity.

To map the gentrification frontier, it will be necessary to find an appropriate indicator of reinvestment and disinvestment, and this involves us in some methodological considerations. The most obvious indicator of the economic turning point associated with gentrification might be a significant and sustained increase in mortgage capital dedicated to building rehabilitation, redevelopment and other forms of neighborhood reinvestment. We already know the critical role of mortgage capital in particular in the geographical division of urban space into recognizable submarkets and in place-specific disinvestment (Harvey 1974; Wolfe *et al.* 1980; Bartelt 1979). Where an adequate flow of mortgage money is not forthcoming, the gentrification of a neighborhood can certainly begin but is unlikely to progress far. Mortgage data have been used widely and to good effect by gentrification researchers (P. Williams 1976, 1978; DeGiovanni 1983; see also Chapters 6 and 7) and represent a very fertile data source in general, but they are not necessarily a sharp indicator of *initial* reinvestment associated with the turning point.

The reason for this is suggested in the introductory epigraph from the Brooklyn developer. Much of the earliest gentrification activity is carried out by developers on the extreme edge of the economic frontier where traditional lenders are generally reluctant to invest yet. In advance of traditional sources, the actual funding mechanisms are diverse, often involve a variety of sources in some form of partnership, and are extremely difficult to trace. One common arrangement of this sort combines several people in a partnership: an architect, a developer, a building manager, a lawyer and a broker. The first three of these will work actively on the conversion of the building while the lawyer handles the gamut of legal matters involved in deed transfer, loan arrangements, state subsidies and tax abatements, and any legal "problems" resulting from the expulsion of existing tenants. The latter is the job of the building manager. All of the partners contribute financially at the beginning of the project, and it is the function of the broker, who may also be the senior contributor, to secure additional private market loans on the basis of this seed money. Where building rehabilitation is organized in this manner, traditional mortgage data would fail to reveal the timing or true dimensions of initial reinvestment.

Other researchers have used state-sponsored programs, such as central government improvement grants in London (Hamnett 1973) or the J-51 program in New York City (Marcuse 1986; Wilson 1985) as a means to date initial reinvestment. While these data have clear utility at a lower scale of analysis, they are even coarser indicators than the flow of mortgage funds, and would be a rather crude means for detecting turning points at the neighborhood scale. A detailed survey of building conditions and an assessment of building deterioration levels might also reveal important information regarding the onset of reinvestment, but it is important to remember that reinvestment may begin substantially in advance of a building's physical upgrading. Indeed, in a detailed survey of displacement pressures in the Lower East Side, DeGiovanni (1987: 32, 35) found strong evidence that physical deterioration may actually be "an integral part of the reinvestment process" as some landlords actually foster adverse physical conditions "to clear buildings of the current tenants" before undertaking major refurbishment or resale. Shifts in the physical condition of a building are better conceived as responses to economic strategies than as causes and therefore provide at best a rough proxy for reinvestment. Finally, building permit data would be ideal, and have been used to good effect in Amsterdam (Cortie *et al.* 1989), but in the US, and especially in New York City, the data are notoriously unreliable. It is widely assumed that as many as a third of inner-city rehabilitations and renovations may take place without permits, and for those that are submitted to the bureaucratic process, the available data are themselves of very uneven quality and difficult to use.

It may well be that tax arrears data provide the most sensitive indicator of initial reinvestment connected with gentrification: the delinquency of the landlord becomes a boon to the researcher. Nonpayment of property taxes by landlords and building owners is one common form of disinvestment in declining neighborhoods. Tax delinquency is in effect an investment strategy since it provides property owners with guaranteed access to capital that would otherwise have been "lost" to tax payments. Insofar as serious delinquency places ownership of the building in jeopardy through the threat of city foreclosure proceedings, we might expect that the extent of tax arrears in a neighborhood would be highly sensitive to reversals in the investment landscape. Where landlords and owners become convinced that substantial reinvestment is possible, they will seek to retain possession of a building whose sale price is expected to increase. Where a building is seriously in arrears, this implies the repayment of at least some back taxes to prevent its foreclosure by the City. Redemption of tax arrears can therefore function as an initial form of reinvestment. This position is supported empirically by Lake's findings in Pittsburgh: among owners of buildings with low to moderate assessed values, there is a clear correlation between their perception of increased property values and their intention to redeem their tax delinquency status (Lake 1979: 192; Sternlieb and Lake 1976).

Salins (1981: 17) makes a similar judgment. He finds tax arrears to be "undoubtedly a very sensitive index of active and incipient housing destruction, especially when viewed in terms of length of delinquency, and the volume of properties at different stages of the arrears foreclosure pipeline." It follows that systematic reversals in arrears level represent an equally sensitive index of reinvestment, yet in neither context have these data been analyzed at a level of geographic disaggregation finer than that of the city or (in New York City) the borough. In addition, as Lake (1979: 207) points out, "Real estate tax delinquency is but a surface manifestation of deep-rooted antagonisms" inherent in the broader processes of urban development. He has in mind the relationship to changing geographical patterns of development but also urban decline and the experience of urban fiscal crisis. Tax delinquency in fact occupies the fulcrum between growth and decline, expansion and contraction, and all that follows from this balance, and the rapid restructuring of the city at the hands of gentrification in the 1970s and 1980s adds further to the pivotal importance of arrearage trends. Peter Marcuse (1984) made the first attempt to use tax arrears data to demonstrate shifts in investment patterns in a neighborhood threatened with gentrification, finding significant evidence of turning points in several census tracts in the Hells Kitchen neighborhood of New York City. The present research builds on this work.

Different cities may have specific procedures for determining tax delinquency and taking into public ownership buildings that surpass a certain threshold of arrearages. Indeed every city has its own political culture of tax delinquency. In most Scandinavian cities, for example, strong state controls and powerful punitive measures make tax delinquency in the residential market a distinct novelty. To a greater or lesser extent, the same is true in most European cities, where national legislation generally governs the property tax systems. In some places, nonpayment of taxes even represents a major crime. In the US, however, the authority to tax is highly decentralized, and different municipalities may have radically different procedures.

In New York City until 1978, foreclosure proceedings could begin against buildings that were twelve or more quarters (three years) in arrears. At the end of a successful foreclosure proceeding, the building was taken into public ownership and placed in "*in rem*" status. A massive wave of residential dis-investment associated with the 1973–1975 recession and the New York City fiscal crisis brought a rapid increase in delinquency rates, and in an attempt to discourage rampant abandonment and disinvestment, then Mayor Beame proposed to make buildings eligible for *in rem* proceedings after only four quarters in arrears. (One- and two-family buildings and condominiums were exempt from this change.) This was enacted into law in 1978. After four quarters in arrears, building owners would have a grace period of a further quarter in which to repay taxes before the onset of *in rem* proceedings. For a further two years, an owner could still redeem the building but at the discretion of the city.

Despite this legislative power, the City Departments of Finance and of Housing Preservation and Development have not customarily initiated proceedings against buildings that are less than twelve quarters in arrears. The procedure is bureaucratic and cumbersome, drawing many complaints from landlords (W. Williams, 1987), but in the 1980s the City administration was clearly reluctant to increase its already large stock of foreclosed and often vacant buildings. It entertains a variety of repayment schemes that encourage building owners to keep title to the building. Many buildings are redeemed well after the twelve-quarter arrears threshold is reached without being foreclosed, others are redeemed according to individual installment plans, and still others are transferred from one owner to another with the purchase price incorporating the redemption of outstanding tax debts. In fact, the City almost immediately regretted the 1978 law, and never seriously implemented the one-year foreclosure requirement (Salins 1981: 18). And by the beginning of the 1990s, with a major economic depression afoot and despite a reduction of City-held housing stock resulting from a major housing rehabilitation program, the Dinkins and Giuliani administrations had both retreated altogether from any aggressive efforts to take buildings *in rem*. During the 1980s, however, despite the fact that the cutoff point for the landlord between "safe" and "unsafe" delinquency levels was not entirely fixed, in practice the administration adhered closely to the prior twelve quarters threshold.

In the postwar period, citywide property tax delinquency levels in New York City peaked in 1976, when over 7 percent of the city's residential buildings were in arrears – an extraordinary figure. After that, however, delinquency levels declined steadily, dropping to a seventy-year low of 2 percent by 1986 (W. Williams 1987), turning up again only at the end of that decade amid another economic depression. In the most immediate sense, the rapid decline in total arrears during the 1980s resulted from the easing of the previous recessions and fiscal crisis of the 1970s (and possibly anticipation of the tighter 1978 law). But mostly it is due to the rapid inflation of real estate prices from the late 1970s. More broadly, declining delinquency levels indicate a very significant reduction in disinvestment levels and a parallel trend toward reinvestment in previously declining real estate. Very few building owners were abandoning properties in the 1980s compared with the 1960s and 1970s. This in turn is directly related to the larger processes of urban restructuring and to gentrification in particular.

Disinvestment is unevenly distributed in the housing market, disproportionately affecting the city's rental stock *vis-à-vis* owner-occupied buildings. This indeed reflects the finding that for many landlords disinvestment is a strategy. As late as 1980, 3.5 per cent of the city's residential properties were in arrears. This figure was dominated by some 330,000 rental apartments, fully 26 percent of the city's entire rental stock (Salins 1981: 17). Overall disinvestment rates declined during the 1980s, but property tax delinquency was increasingly concentrated both geographically and economically in older neighborhoods dominated by large tenement and other multiple-unit rental housing stock. These represent the poorest neighborhoods, destroyed by massive, systematic and sustained disinvestment over a period of three to seven decades: Harlem, the Lower East Side, Bedford Stuyvesant, Brownsville, East New York, the South Bronx.

THE LOWER EAST SIDE

"One must realize," commented a local art critic at the beginning of the area's honeymoon between art and gentrification, "that the East Village or the Lower East Side is more than a geographical location – it is a state of mind" (Moufarrege 1982: 73). Indeed in the 1980s the area was enthusiastically boosted as the newest artistic bohemia in New York City, drawing effusive comparison with the Left Bank or London's Soho. In the gentrification of the Lower East Side, art galleries, dance clubs and studios have been the shock troops of neighborhood reinvestment, although the extraordinary if at times ambivalent complicity of the art scene with the social destruction wrought by gentrification is rarely conceded (but see Deutsche and Ryan 1984). The area was quickly touted as a "neo-frontier," in a deliberate potion of images from the art and gentrification worlds (Levin 1983: 4), and surpassed the staid uptown galleries of Madison Avenue and Fifty-seventh Street, and even the thoroughly corporate art scene of neighboring, once progressive but now utterly gentrified, SoHo. With gushing enthusiasm, the attraction of the Lower East Side was once attributed in the art press to its "unique blend of poverty, punk rock, drugs and arson, Hell's Angels, winos, prostitutes and dilapidated housing that adds up to an adventurous avant-garde setting of considerable cachet" (Robinson and McCormick 1984: 135).

The artistic influx began in the late 1970s and was increasingly institutionalized after 1981 with the widely heralded opening of numerous galleries (Goldstein 1983; Unger 1984) – as many as seventy by the late 1980s. Before the financial shakeout following the 1987 stock market crash on Wall Street, only a couple of miles away, Lower East Side galleries seemed destined for respectability, their prestige enhanced as the establishment's avant-garde. The area also provided the setting as well as the subject of literally dozens of 1980s novels and several movies, including one flirtation with the gentrification genre, Spielberg's *Batteries Not Included*, in which it takes space aliens to rescue Lower East Side victims of gentrification-induced displacement. But the romanticization of poverty and deprivation – the area's "unique blend" – is always limited, and the neon and pastiche sparkle of aesthetic ultra-chic only partly camouflages the harsher realities of displacement, homelessness, unemployment and deprivation in a neighborhood converted into a new frontier by "the fine art of gentrification" (Deutsche and Ryan 1984).

Bounded by Fourteenth Street to the north, the Bowery to the west and the East River Drive to the south and east, the Lower East Side sits to the east of Greenwich Village and SoHo and north of Chinatown and the financial district (Figure 1.1). The area comprises Community Board 3 in Manhattan. Except for several public housing complexes on the easternmost edge of the area, the housing stock of the Lower East Side is dominated by four- to six-storey "railway" and "dumbbell" (old law) tenements built in the late nineteenth century and now either heavily deteriorated following decades of disinvestment or else recently restored and roughly polished by gentrification. These are interspersed with occasional public housing complexes of ten or more storeys constructed in the immediate postwar period, a few apartment blocks built in the earlier part of the twentieth century, and a few older tenements and town houses that predate the second half of the nineteenth century. Socially even more than physically, the Lower East Side of the 1980s was a crazy quilt of yuppies and punk culture, Polish and Puerto Ricans residents, Ukrainian and African-American working class, quiche and fern restaurants and homeless shelters, surviving ethnic churches and burned-out buildings. From the turn of the century till after World War II, it was not only the preeminent reception community for European immigrants to the United States, but an area of intense socialist, communist, Trotskyist and anarchist organizing, a major progenitor of New York intellectuals and at the same time an extraordinary seedbed of small entrepreneurs and businesses. This exceptionally variegated history and geography would encourage us to believe that to the extent that recognizable geographical patterns of reinvestment can be found here, they are redolent of deeper regularities in the gentrification process.

Like that of most poor areas in the city, the population of the Lower East Side declined dramatically in the 1970s, followed by increasing stabilization in the 1980s. If the 1980 population had decreased by over 30 per cent in the preceding ten years, rents nonetheless increased by between 128 per cent and 172 per cent in the area's census tracts – universally higher than the citywide increase over the same period of 125 percent. Fully a quarter of all households were below the poverty line in 1980, but there was considerable variation by census tract (from 14.9 per cent to 64.9 percent) (US Department of Commerce 1972, 1983). Population decline in the 1970s was not repeated in the 1980s. According to data from Con Edison, the local gas and electricity company, which maintains rather accurate records on who is using its gas and electricity, the number of household utility hook-ups in the area fell precipitously until 1982 when it began to rise again, presumably as a result of gentrification. By 1990 the census counted 161,617 residents, a 4.3 percent increase over the 1980s.

To turn to disinvestment, the peak years of total tax arrears in the Lower East Side came in 1976 and 1977, but by 1986 the level of arrearages had dropped by fully 50 percent, and continued to drop until 1988. Figure 9.1 contrasts the history of disinvestment (total private market residential tax arrears) and population levels (Con Edison accounts) on an annual basis between 1974 and 1986.[1] It indicates that although some redemption of tax arrears began after 1976, it was not until after 1982 that this reinvestment manifested itself demographically in a rise in the number of households. Table 9.1 provides parallel data on vacancy rates for the northern East Village area. All of the census tracts in the area experienced peak vacancy rates between 1976 and 1978, and by 1984 virtually every tract had more than halved its vacancy rate. This suggests a lag of perhaps as much as six years between earliest reinvestment and repopulation, and offers further empirical support to the thesis that economic shifts lead demographic change in the gentrification process.

MAPPING GENTRIFICATION

As Figure 9.1 begins to suggest, the decline in disinvestment after 1976, following the amelioration of the fiscal crisis and economic recession, initiated a consequent reinvestment in the Lower East Side that was sustained even through the recession of the early 1980s, fading only with the depression that began at the end of that decade. The data demonstrate considerable internal differences geographically and historically in the disinvestment and reinvestment processes, however, and from this we can begin to compile a picture of the "frontier of profitability" connected with gentrification.

Table 9.1 Level and year of peak vacancy in the northern East Village (New York) area

Rank	Tract	Highest % unoccupied	Year	% unoccupied 1984
1	26.01	25.06	1978	11.87
2	26.02	24.87	1978	10.53
3	22.02	21.11	1976	11.48
4	28.0	20.78	1976	5.59
5	34.0	16.24	1978	6.40
6	30.02	11.77	1976	5.20
7	36.02	11.77	1976	5.20
8	32.0	9.27	1976	3.77
9	38.0	6.98	1976	3.30
10	40	5.77	1976	2.88
11	42	4.50	1976	1.49

Source: MISLAND, Con Edison File

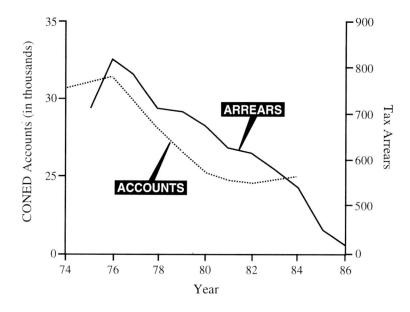

Figure 9.1 Disinvestment (arrears) and population change (accounts) in the Lower
East Side, New York, 1974–1986

Tax arrears data are collected by the Department of Finance and are
available on the City's MISLAND database. For each census tract, a summary is
available showing the extent of arrears. Tax lots in arrears can be categorised
according to the severity of delinquency: low (3 to 5 quarters in arrears), intermediate (6 to 11 quarters), and serious (12 or more quarters in arrears).
Table 9.2 provides data on serious residential tax arrears for the East Village
section (north of Houston Street) between 1975 and 1984. Given the *de
facto* threshold of 12 quarters before the initiation of *in rem* proceedings, low
levels of arrears (3–5 quarters) tend not to reflect sharp and significant
changes in delinquency and redemption. However, arrears in the intermediate
(6–12 quarters) and serious (12+ quarters) categories do reveal an interesting
historical trend (Figure 9.2). Until 1980 there is a clear inverse relationship
between the number of buildings in the intermediate category and those in
serious delinquency. Building owners would seem to have followed an obvious
strategy: buildings are partially redeemed in 1978 and 1979 (bringing them
back from the serious to the intermediate arrears category), presumably
to avert the threat of foreclosure at the hands of a 1978 citywide vesting as
well as the new delinquency law. In 1980 a new wave of delinquency began
as about 170 properties slipped back from intermediate to serious delinquency.
After 1980, however, the inverse relationship between serious and intermediate delinquency is suspended as both categories decline significantly.
Despite the national recession of 1980–1982, which seriously curtailed
residential construction, the decline in disinvestment continued unabated
until the end of the 1980s. Only with the threat of a major foreclosure and
vesting proceedings in 1985 is there a repeat of the inverse relationship

Table 9.2 Trends in residential tax arrears in the East Village, New York

Census tract	Number of quarters in arrears	Number of tax lots in arrears											
		1975	1976	1977	1978	1979	1980	1981	1982	1983	1984	1985	1986
All	+12	241	369	402	344	244	417	385	352	338	324	79	107
22.02	+12	16	24	34	27	12	33	34	30	28	28	8	9
26.01	+12	39	72	87	66	53	88	76	71	73	73	10	11
26.02	+12	37	67	74	80	50	96	77	72	71	72	15	17
28	+12	40	57	64	49	34	50	44	43	42	38	5	14
30.02	+12	11	12	19	15	9	16	21	22	20	16	3	4
32	+12	22	27	30	26	24	44	44	37	35	39	19	23
34	+12	32	54	49	46	31	54	52	42	42	39	8	17
36.02	+12	14	18	16	15	11	14	14	12	10	8	8	3
38	+12	20	26	22	18	16	17	15	14	10	6	5	6
40	+12	9	12	7	2	4	5	8	9	7	5	1	3
42	+12	1	0	0	0	0	0	0	0	0	0	0	0

Source: Manhattan MISLAND Report: New York Department of City Planning; New York Property Transaction File; Real Property File

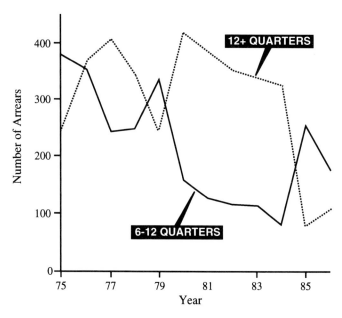

Figure 9.2 Cycles of serious and intermediate disinvestment in the Lower East Side, 1975–1986

between serious and intermediate delinquency, but this is a ripple within a larger decline in overall disinvestment. This suggests that serious reinvestment in the area as a whole began after 1980, the year of peak levels in serious delinquency.

Corroborative evidence comes from sale price data. Whereas median per unit sale prices for the whole Lower East Side rose only 43.8 per cent between 1968 and 1979 (a period in which the inflation rate exceeded 100%), prices between 1979 and 1984 rose 146.4 per cent (3.7 times the rate of inflation) (DeGiovanni 1987, 27).

In order to disaggregate the arrears data geographically and to identify the gentrification frontier it is necessary first to identify a "turning point" for each census tract. The turning point represents the year of peak serious arrears for each tract. Figure 9.3 provides four illustrations of this procedure. In Figures 9.3a, 9.3b and 9.3d, the turning points are 1980, 1982 and 1976 respectively. Where tracts exhibit a bimodal distribution, as in tract 34, for example (Figure 9.3c), the later peak was identified as the turning point, since our concern here is the date of reversal from disinvestment to sustained reinvestment.

The inverse relationship between intermediate and serious arrears also seems to be continued in this period, although for reasons of space the intermediate data are not graphed here. In every case, except census tract 42, where the number of intermediate arrears is too low to allow reasonable statistical comparison, intermediate arrears levels peak prior to the turning points in serious delinquency; peaking intermediate arrears always preceded the turning point in serious arrears. Eight tracts had an intermediate peak in

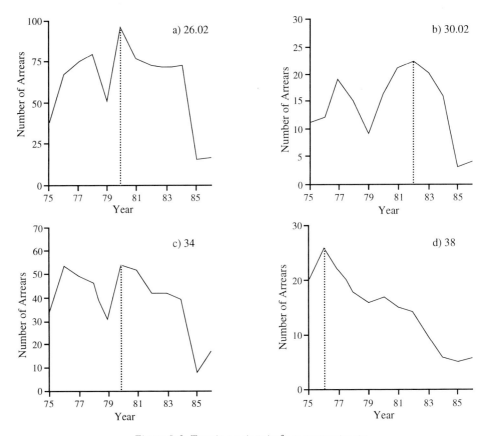

Figure 9.3 Turning points in four census tracts

1975–1976 with the remaining two, both in the southeast of the area, peaking later, in 1979. This would seem to reaffirm the earlier interpretation whereby, during the 1970s, many properties hovered around the 12+ quarters threshold. In fact they went through a cycle of intermediate delinquency followed by serious delinquency then redemption again to intermediate status. The timing of this cycle was geared to the timing of the City's foreclosure proceedings.

For purposes of mapping the gentrification frontier, data were analyzed from the twenty-seven census tracts with private housing in the entire Lower East Side. The earliest turning points were generally on the western edge of the area, coming in 1975–1976; the latest lay well to the east and occurred in 1983–1985. Every tract had experienced a turning point by 1985. The resulting geographical pattern of turning points demonstrates extreme statistical autocorelation. The turning points were then mapped and the data were generalized via a least-squares method into a chorographic map of the development of gentrification.[2] In Figure 9.4, annual contour lines join points with the same chronological turning points, and the shading highlights the different stages of the onset of reinvestment. By way of interpretation,

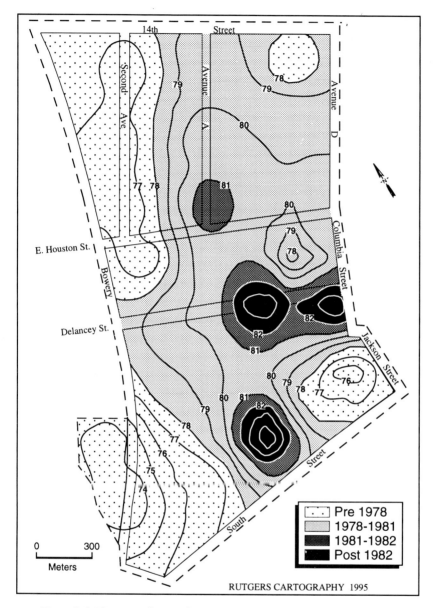

Figure 9.4 The gentrification frontier in the Lower East Side, 1974–1986

where significant space intervenes between chronological contour lines, reinvestment is diffusing rapidly; steep contour slopes indicate significant barriers to reinvestment. The frontier is most evident where there are no enclosed contours (that is, no peaks or sink holes). Peaks, with later years at the center of enclosed contours, represent areas of greatest resistance to re-investment while sink holes, with earlier years at the center, represent areas opened up to reinvestment in advance of surrounding areas. The major pattern that emerged is a reasonably well-defined west-to-east frontier line with the earliest encroachment in the northwest and southwest sections of the Lower East Side. The reinvestment frontier pushes east until obstructed and slowed by localized barriers in the east and southeast.

In geographical context, the gentrification frontier clearly advanced eastward into the Lower East Side from Greenwich Village, SoHo, Chinatown and the financial district. Greenwich Village has always been a bohemian neighborhood, but after some decline in the 1930s and 1940s it began in the 1950s and 1960s to experience an early gentrification. SoHo's gentrification came a few years later but was essentially complete by the late 1970s. In Chinatown, an influx of Taiwanese capital in the mid-1970s and a later influx by Hong Kong capital provided the means for a rapid northward and eastward expansion of Chinatown, only some of which might properly be described as gentrification. In addition, it is worth pointing out that the extent of pre-1975 reinvestment on the border of Chinatown to the south (Figure 9.4) may be exaggerated owing to border effects in the statistical analysis.

Barriers to the advancement of the frontier are apparent in several localized peaks, especially on east Delancey Street and on the southern edge near South Street. Delancey Street is largely commercial, a wide thoroughfare leading to the Williamsburg Bridge which connects Manhattan and Brooklyn, and its congestion, noise and impassability may well have hindered reinvestment. More generally, these peaks can be interpreted as demonstrating the limits of gentrification: the eastern and southern edges of the area are fringed by large public housing projects that could be expected to act as significant barriers. In addition, these nodes of resistance to reinvestment coincide with the traditional heart of the Lower East Side, where disinvestment continued until as late as 1985, well into the 1980s economic recovery. This is also the poorest area, the last stronghold of a Latino population, and the focus of "Operation Pressure Point," a gentrification-induced police crackdown on street drugs that began in 1985 – the year of the final turning point, as it happens. Figure 9.5 shows an alternative three-dimensional pictorial of the "gentrification surface," etched as the frontier courses through the neighborhood. The lowest areas experience reinvestment first, the highest later. Gentrification here flows uphill, as it were, against disinvestment.

There seem to have been two distinct periods of reinvestment in the Lower East Side housing stock. The first took place between 1977 and 1979, especially in the western and northern blocks, and in the period after 1980. The second phase of reinvestment encompassed the southern and eastern blocks in addition to those already recapitalized in the earlier phase. While it is important to bear in mind that reinvestment in the form of tax arrears redemption does not necessarily imply the kind of productive reinvestment in building refurbishment and redevelopment that betokens gentrification and urban restructuring, and might only indicate a speculative market, the reinvestment in the western and northern blocks in the late 1970s does seem to have been sufficiently sustained to prevent a major recurrence of disinvestment in the recession of 1980 to 1982. And indeed, reinvestment in the area to the west of First Avenue has been longer in duration, more sustained, and

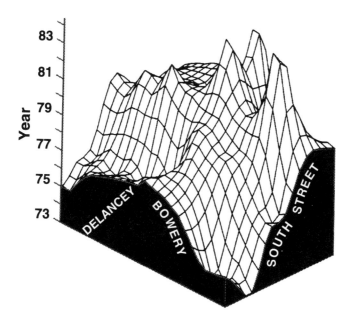

Figure 9.5 The Lower East Side as gentrification surface

more broadly based: the census tracts in the east and southeast sector sustained the highest population losses (58–74 per cent) throughout the 1970s and began an economic upturn only well into the 1980s.

Far from "slouching toward Avenue D," as art critics Walter Robinson and Carlo McCormick (1984) put it, gentrification moved eastward through the neighborhood at a fair clip. From 1975 to 1981, the profit frontier moved at an average speed of between 100 and 200 meters per year. It is important to add the caveat that this figure represents an averaging across a period in which the market was highly volatile: high disinvestment in 1976–1977 presaged a slow diffusion of the profit frontier, but was followed by a more rapid shift until 1980, when massive disinvestment again slowed the diffusion of the reinvestment frontier until 1982. Further, these data cover only two short cycles of reinvestment and disinvestment, and care is necessary in generalizing the conclusions. The diffusion of the "frontier of profitability" in gentrifying neighborhoods is of course sensitive to external economic and political forces, and has its own block-by-block microgeography; it may be a stop-and-go process as much as a smooth progression.

This pattern of reinvestment in the Lower East Side coincides closely with local observations. As the Real Estate section of the Sunday *New York Times* enthused, "Gentrification continued its inexorable march across 'Alphabet City' " – from Avenue A to Avenue B, then C and D (Foderaro, 1987). To many observers, gentrification seemed to encroach from the western border of the East Village, with the more established Greenwich Village providing the adjacent impetus. The area to the north of Fourteenth Street, including Gramercy Park but also Union Square farther to the west, has been the target

Plate 9.2 "See no homeless" in the new urban reservation

of early redevelopment activity, leading to the construction in 1987 of the Zeckendorf Tower Condominium, which anchors the northwest corner of a gentrifying Lower East Side. And although Stuyvesant Town (a moderate- and middle-income high-rise complex on the north side of Fourteenth Street farther to the east) may initially have hindered the southward diffusion of higher land values, once the process began it may equally have acted as a northern stabilizer to the gentrification of the East Village. By contrast, the southern and eastern blocks experienced a deeper disinvestment, as suggested by the inordinately high peak vacancy rates (Table 9.1). South of Houston Street, where housing conditions are the worst in Manhattan outside Harlem, reinvestment was generally more tardy. Thus reinvestment in the Lower East Side began not in the area of deepest disinvestment and abandonment but on the borders (Marcuse 1986: 166). It was in these border areas that a killing could be made, so to speak, with little risk of simultaneously being scalped.

CONCLUSION

This analysis of reinvestment turning points and map of the frontier of profitability are highly suggestive. The local complexity of the pattern and its deviation from a straight-line diffusion process should come as no surprise; indeed Frederick Jackson Turner, author of the nineteenth-century "end-of-the-frontier" thesis, was challenged on exactly this point, namely that while the larger frontier line may have swept through, it left behind resilient pockets of frontier existence. As with the original frontier, the gentrification line is not so much a "wall" of equal and continuous development as a highly uneven and differentiated process. Nonetheless, mapping the gentrification frontier helps significantly to demystify the frontier language through which so much gentrification has been interpreted in the popular press and help us discern the economic geography of urban change that gives this language a semblance of reality. If this mapping detects the economic frontier of gentrification, the political culture and the cultural politics of the frontier present a very different tapestry.

FROM GENTRIFICATION TO THE REVANCHIST CITY

After the stretch-limo optimism of the 1980s was rear-ended in the financial crash of 1987, then totaled by the onset of economic depression two years later, real estate agents and urban commentators quickly began deploying the language of "*de*gentrification" to represent the apparent reversal of urban change in the 1990s. "With the realty boom gone bust in once gentrifying neighborhoods," writes one newspaper reporter,

> co-op converters and speculators who worked the streets and avenues . . . have fallen on hard times. That, in turn, has left some residents complaining of poor security and shoddy maintenance, while others are unable to sell their once-pricey apartments in buildings where a bank foreclosed on a converter.

"Degentrification," explains one New York realtor, "is a reversal of the gentri-fication process": in the 1990s, unlike the 1980s, "there is no demand for pioneering, transitional, recently discovered locations." Those few real estate deals that are transacted, he suggests, have retrenched to "prime areas" (quoted in Bagli 1991). "In the 1970s, the theory was that a few gentrified areas would have a contagious effect and pull up neighboring districts," but "that didn't happen," says another commentator. Most bluntly, in the words of Census Bureau demographer Larry Long, "gentrification has come and gone" (quoted in Uzelac 1991).

Such media proclamations of the end of gentrification have begun to find broader support in the academic literature, where commentators were in any case usually more bromidic in their rhetoric about gentrification. In a clearly argued essay drawing on a Canadian case study, Larry Bourne anticipates "the demise of gentrification" in those few cities where, he suggests, it had even a minor significance in the 1980s. Gentrification "will be of less importance as a spatial expression of social change during the 1990s than it has been in the recent past" (Bourne 1993: 103). The last decade and a half, he suggests, were

> a unique period in post-war urban development in North America – a period that combined the baby boom, rising educational levels, a rapid growth in service employment and real income, high rates of house-hold formation, housing stock appreciation, public sector largesse,

widespread (and speculative) private investment in the built environment, and high levels of foreign immigration. This set of circumstances, except for the latter, no longer prevails.

(Bourne 1993: 105–106)

The "postgentrification era" will experience a much reduced "rate and impact of gentrification" in favor of a more unevenly developed, polarized and segregated city.

The coining of "degentrification" and the prediction of gentrification's demise are part of a wider "discourse of urban decline" (Beauregard 1993) that has repossessed the public representation of urbanism in the 1990s, especially in the US. Historically, according to Beauregard, this discourse of decline has been "more than the objective reporting of an uncontestable reality"; rather, the discourse "functions ideologically to shape our attention, provide reasons for how we should react in response, and convey a comprehensible, compelling, and reassuring story of the fate of the twentieth-century city in the United States" (1993: xi). The recrudescence of this discourse in the 1990s has been dramatic. Gone is the white, upper-middle-class optimism of gentrification which was supposed to reclaim the "new urban frontier" in the name of largely white "pioneers," an optimism that significantly modulated the discourse of decline during the 1970s and especially the 1980s. In its place has come the revanchist city.

THE REVANCHIST CITY

In the 1990s an unabated litany of crime and violence, drugs and unemployment, immigration and depravity – all laced through with terror – now scripts an unabashed revanchism of the city. The revanchists in late nineteenth-century France initiated a revengeful and reactionary campaign against the French people (Rutkoff 1981), and it provides the most fitting historical pretext for the current American urbanism. This revanchist antiurbanism represents a reaction against the supposed "theft" of the city, a desperate defense of a challenged phalanx of privileges, cloaked in the populist language of civic morality, family values and neighborhood security. More than anything the revanchist city expresses a race/class/gender terror felt by middle- and ruling-class whites who are suddenly stuck in place by a ravaged property market, the threat and reality of unemployment, the decimation of social services, and the emergence of minority and immigrant groups, as well as women, as powerful urban actors. It portends a vicious reaction against minorities, the working class, homeless people, the unemployed, women, gays and lesbians, immigrants. The revanchist city is screamingly reaffirmed by television programming. The "gentrification of prime time" (B. Williams 1988: 107) in the 1980s has given way to an obsessive portrayal of the apparent danger and violence of everyday life. The local news, "Cops," "Hard Copy," "911," a whole cable channel devoted to "Court TV," together with talk radio, militia radio and late night-crueltymongers like Rush Limbaugh all blend prurience and revenge as an antidote to insecure identities. Sixteen months of daily coverage of the O. J.

Simpson trial, and his eventual acquittal, only hardened the racial topologies of vengeful reaction, while the real message was that class and money were powerful enough to supplant race, and women were the all-round losers. So extreme is the desire for revenge, apparently, that a group of lawyers and investors has formed in California with the express purpose of airing on pay-TV an "execution of the month" ("Production Group . . . " 1994).

The revanchist city represents a reaction to an urbanism defined by recurrent waves of unremitting danger and brutality fueled by venal and uncontrolled passion. It is a place, in fact, where the reproduction of social relations has gone stupifyingly wrong (Katz 1991c), but where the response is a virulent reassertion of many of the same oppressions and prescriptions that created the problem in the first place. "In the U.S.," says Gilmore, quoting Amiri Baraka, "where real and imagined social relations are expressed most rigidly in race/gender hierarchies, the 'reproduction' is really a *production* and its by-products, fear and fury, are in service of a 'changing same': the apartheid local of American nationalism" (Gilmore 1993: 26).

Third World cities have for a long time been scripted in the West as similar kinds of "revanchist cities," cities where nature and humanity habitually take vicious revenge on a degenerate and profligate populace. The eco-reactionary and eugenicist language of the "teaming masses" or the population bomb – resurgent again today – is interwoven with a subtext of just deserts: the plagues, earthquakes and human massacres that define these cities in the Western press are presented as a revenge of nature (human or otherwise) against a fatally flawed segment of humanity. The organized murder of street kids in Rio de Janeiro, the Hindu massacres of Muslims in Bombay, the pre-election slaughter of South Africans in Durban (passed off as tribal warfare, but fueled by the South African security forces), the mayhem in Baghdad streets after the barbaric US bombing in 1991 and again in 1993, and the spectacular violence in Rwanda – these and many other dramas present Third World cities to Western audiences not simply as places of extraordinary and often inexplicable violence but as places of inherently revengeful, perhaps lamentable, but often justifiable violence. The revanchist city of the 1990s, however, is more about the rediscovery of enemies within.

In retrospect, the emerging revanchist city may have been most drama-tically announced by the publication in 1987 of Tom Wolfe's visceral portrait of New York, *Bonfire of the Vanities*. This book, and the Hollywood screen version that followed, is the story of the decline and fall of an erstwhile "master of the universe" – a Wall Street trader whose world implodes. In a tale that took considerable liberty with nonetheless recognizable events and characters in the radically transformed New York of the late 1980s, the fictional Sherman McCoy, a Park Avenue scion of upper-class WASP respectability, becomes a seemingly unwitting victim to a world that he felt had been stolen from him and his class. Following a car accident in the remote Bronx in which a black teenager is knocked over and eventually dies, McCoy is brought face to face with a world of immigrants, newly powerful minority politicians and preachers, and the Kafkaesque legal bureaucracy of the Bronx Courthouse. Despite his class power and connections and despite his millions – perhaps because of them, Wolfe allows – he is unable to extricate himself from the fall-out from a crime that he didn't even commit. Wolfe reserves special loathing for the African-American reverend of a Harlem church, easily recognizable as modeled on a real person, and however sardonic his indictment of McCoy's class pomposity, *Bonfire of the Vanities* is nevertheless the tale of a white upper-class male, unreasonably victimized by a world he no longer completely controls.

The last few years have thrown up many variants of the revanchist city. A Scandinavian airlines magazine features Miami writer Carl Hiaasen, whose crime novels paint a lurid picture of "Miami vice." For Hiaasen, himself a second-generation Norwegian immigrant, the origins of Miami vice lie directly with overpopulation:

> we are never going to save this place until we get some breathing room, until this population is of manageable enough size. This is not an ecology which supports four million souls. There is not enough water, there is not enough land. We have bunched together so tightly that it is now erupting in this horrible violent crime. It takes the form of maybe a race riot one summer; it may just be the random homicide rate.
>
> I doubt that many people who moved here ten or fifteen years ago dreamt that they would have to put bars on their windows or carry Mace to the grocery store or worry about being car-jacked or mugged on the way home from the airport. It's all the result of having too damned many people.
>
> (quoted in Rudbeck 1994: 55)

If largely minority immigrants – from Haiti, Cuba, Colombia and elsewhere in the Caribbean and Latin America – are Hiaasen's primary target, he is no consistent racist: "As far as I'm concerned," he says, referring to Marco Island a Gulf resort that attracts European, Canadian and Midwestern visitors as well as Caribbeans and Latin Americans, "that would be a great place for a tactical nuclear strike" (p. 54).

Such a hysterical causal connection between crime, immigration and "overpopulation" may make for good tabloid copy, but it is just as certainly bad science. Crime in particular has become a central marker of the revanchist city, the more so as the fears and realities of crime are desynchronized. "Crime surpasses healthcare and 'the economy' as current public anxiety number one," suggests Ruth Wilson Gilmore,

> even though it is well reported that in recent years average crime rates have gone down. In the contemporary U.S., crime constitutes a double displacement: First, it is symptomatic of the disorder of people's lives when wage-money is hard to come by. . . . Second, it organizes people's fears brought about by the vertigo of economic insecurity by identifying an enemy whose vanquishment will restore security.
>
> (Gilmore 1994: 3; see also Ekland-Olson *et al.* 1992)

Two events on different coasts, equally coded by race and nationalism entwined with class and gender, crystallized the emerging revanchism of the

American city at the dawn of the 1990s. In Los Angeles, widely heralded in the 1980s as the new, raw, Pacific urbanism for a new century, the 1991 uprising following the acquittal of four police officers in the vicious beating of Rodney King defied habitual media efforts to explain the "riot" as a simple black assault on whites. The flood of racial stereotypes as a means to explain the uprising was deafening and in the end unsuccessful, for it was, as Mike Davis put it, "an extremely hybrid uprising, possibly the first multi-ethnic rioting in modern American uprising" (Davis in Katz and Smith 1992: 19; see also Gooding-Williams 1993). Likewise, the bombing less than a year later of New York City's World Trade Center – simultaneously a symbol of 1970s downtown renewal (and the massive displacement this involved) and the 1980s global urbanism – evoked vivid images of a real-life *Towering Inferno*, and unleashed a xenophobic media hunt for "foreign Arab terrorists" (Ross 1994). While the complete failure of the building's security systems led to its depiction as a "sick building" in a "sick city," the Trade Center bombing cemented the connection between American urban life and apparently arbitrary but brutal violence (terror) on the international scene. Even the usually astute critic Paul Virilio fell into line with his assessment that the Trade Center bombing "inaugurates a new age of terrorism" – an "age of disequilibrium" surpassing a past and presumably preferable "equilibrium of terror" (Virilio 1994: 62). (It is difficult to avoid the aside here that we used to level at neoclassical economics, namely that the difference between equilibrium and disequilibrium depends an awful lot on where you stand.) In any case, the xenophobic hysteria that followed enlisted even the *New York Times*, whose language of blithe exaggeration passed for uncontested fact as it documented the search for foreign conspira- tors – "a ring accused of plotting to blow up New York City" (Blumenthal 1994). No mere Manhattan Project that.

Two later events further solidified not just the emerging revanchism of the US city but the inevitably international context in which this revanchism is being fashioned. When Dr. Baruch Goldstein, the American Jewish settler from the West Bank, sprayed the Hebron mosque with machine-gun fire on February 25, 1994, assassinating twenty-nine Palestinians who had been attending Ramadan prayers, the *New York Times* responded by exploring the emotional turmoil and embarrassment felt by many Israelis about the massacre (Blumenthal1994). Very little of a systemic nature was diagnosed from the event, which was broadly laid to Goldstein's "instability," an un- fortunate psychology. For the *Times* and for most of its US audience, the murdered Palestinians, by contrast, remained unnamed and presumably therefore unimportant; only as a belated afterthought were their names even reported in a few outlets of the US press. They were uncounted (literally: for days, reported estimates of casualties ran from 22 to 43, and only after weeks did the US press desultorily settle on an official figure of 29).

When, less than a week later, a gunman in broad daylight shot up a van of orthodox Lubavitch Jews on the Brooklyn Bridge, killing one person, the pattern of response was in one sense astonishingly similar, despite the fact that the episode represented a diametrically opposed violence. The adminis- tration of Mayor Giuliani and the city's media again focused on the "rage and pain" of New York's orthodox Jews. They speculated, or denied speculation, that "the attack might have been only a hurried act of vengeance for the slaying of dozens of Muslims in Hebron." Such speculation was increased when the NYPD arrested Rashad Baz, whom the *New York Times* identified as "an alien" – actually a Lebanese citizen whose visa had run out – and charged him with homicide. Subsequently no such connection was established, in fact, but the innuendo of Baz's supposed alienness made the link anyway: "his possessions included Islamic prayer beads and other religious articles, as well as a newspaper clipping about a bombing in Lebanon," reported the *New York Times*. Lebanon, Hebron: what's the difference?

This case was immediately scripted as a *national atrocity*. Both words are important. Had it happened at night in the nearby Brooklyn neighborhood of Bedford Stuyvesant it would most likely have been logged in police computers as just another *local* "drive-by shooting." And if it had involved African-Americans on both sides, it might not even have merited a mention in the *Times*, never mind national attention as "atrocity": one more ghetto murder. A little more coverage and angst would have been expended if the parties could have been passed off as assimilated white Americans, especially if the victim (or the shooter) was visiting the city from respectable upper-middle-class suburbs somewhere else. So what made the Brooklyn Bridge killing so symbolic? Apart from suggesting that international political struggles were at home on the streets of New York as much as Beirut, the shooting confirmed an already ubiquitous rescripting of internal "enemies" – Arab immigrants – as really external. Second, this attack on Lubavitchers became an immediate means by which Jews more generally could be reinstated as victims, countering the discordant impact of the Hebron massacre.

Comparison was quickly and widely made between this case and the Crown Heights case two years ago when a young black child was killed by a Lubavitch youth who ran a red light; in an ensuing riot a young Australian Lubavitcher was killed. In that case, the local black population widely blamed the police for being more concerned to protect the car driver than to get the child to hospital, while Lubavitch sect members accused the police of deliberately not quelling the riot that ensued. The latter accusations were aimed at the then mayor, David Dinkins, and the police chief, and became a central issue in the 1993 mayoral campaign. That campaign elected a new mayor, Rudy Giuliani, who happens to be white and who is the first Republican mayor since Robert Lindsay in the late 1960s, and who has spearheaded a particularly revanchist urban politics in the mid-1990s.

The second case represents an obverse of a different sort: the bombing of the Oklahoma City Federal Building on April 19, 1995. The massive explosion, which killed 168 people, was quickly billed by CNN as "Terror in the Heartland," as if to point out that it was not only in New York and Los Angeles that Americans were vulnerable to international terrorism. In the hours following the blast, the FBI organized a massive manhunt for "two Middle-Eastern men" who were supposedly witnessed running from the scene. The media found various obliging "experts" to declare that the bombing bore all the hallmarks of "Middle Eastern terrorism" and used this testimony to spin a bewildering array of conspiratorial scenarios. Muslim Americans were harassed, the Nation of Islam was fingered as possible fundamentalist perpetrators from within, and two men who had driven to Oklahoma in hopes of a quicker expedition of immigration documents from the Immigration and Nationalization Service (INS) were detained. A young man who had just got off a plane from London, and who was of "Middle Eastern descent," was held for several days for being in the wrong place at the wrong time.

The anti-Semitism and racism of this response were only outstripped by that of the stunned reaction to the news several days later that Timothy McVeigh, a European-American, right-wing extremist, ex-Army cadet, and sometime antigovernment militia associate, had been detained as a suspect. A second suspect was apprehended, and despite desperate speculation that they had been used and that "Middle Eastern terrorists" were still ultimately to blame, it was increasingly obvious that responsibility for the Oklahoma City bombing lay in the Middle West not the Middle East, and among white boys to boot; "Terror in the Heartland" took on a momentarily more sinister meaning. For an instant, indeed, the meaning seemed inassimilable. To many, from Oklahoma to Washington, DC, the idea that it was a "homegrown American kid" who was responsible for what was widely billed as the worst act of terrorism in US history was, if anything, worse. That "foreign Arabs" might hate America enough to commit a bombing was understandable, they seemed to be saying (in a moment of inadvertent self-revelation), but for it to be "one of our own," as many officials, journalists and interviewees blurted out, was unfathomable. The discourse changed overnight to the psychology of militia members and the irrationality of antigovernment attacks, even as the Republican majority in Congress, which was leading its own vicious attack on poor and working-class Americans, women, minorities and immigrants, via a particularly nasty antigovernment rhetoric, were peculiarly silent. They recouped themselves only later with well-publicized hearings about deadly government violence at Waco (against the Branch Dravidians) and against white supremacists at Ruby Ridge.

There are obvious questions about this pervasive framing of the Oklahoma City bombing. What gets to count as "terrorism," and what is forgotten: did slavery and lynching not account for a more brutal and more protracted terrorism in US history? And who gets to count as "our own": would the scripting of "us" and "them" have been different had the suspects been black instead of white? But beyond this, even the most cynical commentator probably could not have predicted the legislative course of a Congress suddenly galvanized against terrorism. The antiterrorist legislation presented by the Clinton administration in response to Oklahoma City (and enthusiastically taken up by Congress) certainly included sweeping provisions restoring wide pre-1970s authority for FBI surveillance. But even more precious to the administration and legislators were the widespread provisions against "foreign terrorism"; among other provisions, the US government sought the power to designate (virtually at will) certain "foreign" organizations as inherently terrorist, and to criminalize both membership in and financial support for any such organization by US citizens. The message was clear: sure, domestic terrorism may actually have been responsible for the murders in Oklahoma City, but that was an aberration; foreign terrorists are the real threat and they are therefore the appropriate target of new antiterrorist legislation.

We might well be reminded of Menachem Begin, actually, who is reputed to have said, picking up on some anti-Jewish responses in the US to the bombing of the US Marine compound in Beirut in 1983: "goyim kill goyim, and still they blame the Jews!" In Oklahoma City, Americans killed Americans, and still they blamed the "foreigners."

The revengeful reaction to the city in the 1990s represents a response to a failed urban optimism at the end of the 1980s. For many who succeeded as yuppies in the previous decade, the 1990s has been a time of economic retreat and the dismal defeat of often unrealistic aspirations. For the majority for whom yuppiedom was always beyond reach but nonetheless made a very tangible desire, the despair was if anything more real with the onset of economic depression between 1988 and 1992. The economic depression not only affected jobs and wages but also deflated the real estate industry, which not only led much of the economic boom but became a central symbol of its upward spiral. Gentrification in particular and the urban fast life more generally came to symbolize 1980s yuppie aspirations much as suburban domesticity did in the postwar boom.

These are not new themes, of course. Antiurbanism runs deep in US public culture (White and White 1977), and the postwar portrayal of the city as jungle and wilderness – the Wild City, as Castells (1976) put it – was never entirely absent through the 1980s, accompanying as much as contradicting the redemptive gentrification narrative. What *is* new is the extent to which this panoply of "fear and fury" (Gilmore 1993: 26) has again come to monopolize public media visions of urban life, and the extent to which the revanchist US city is now recognized as an inherently international artifact. From NAFTA to the World Trade Center, the safety of US borders, real and imagined, has dissolved. Not since the villainization of the city in the 1910s and early 1920s of this century, when European immigrant socialists were identified as attacking the fabric of urban democracy, has US antiurbanism involved such an explicitly international recognition. Neither the seeming *deus ex machina* of nuclear attack nor the McCarthyism of the cold war produced comparable visions of a US urbanism vulnerable to foreign attack from within; and for their part, the civil rights uprisings of the 1960s, which had an effect on urban structure sufficient to provoke the racist term "white flight," was represented as a largely domestic question, connections to the anti-Vietnam War movement notwithstanding.

What *is* surprising, perhaps, is not so much that a new antiurbanism incorporates a reluctant acknowledgment of the internationalization of local social economies in the last two decades. Rather what is surprising is that media self-representations of US cities – ostensibly among the most cosmopolitan of cities, at least in terms of the flow of capital and culture, commodities and information – were so systematically able to insulate and indeed isolate the triumphs and crises of American urban life from international events in general but especially from the results of US military, political and economic policy abroad. It is hardly an exaggeration to say that the internationalism of the US city was largely restricted on the one hand to recognizing the connections of capital and the market and on the other to the recognition of nostalgic if palpably real Little Italys, Little Taiwans, Little Jamaicas, Little San Juans that dotted the urban landscape, as if to allow a tokenist internationalism at the neighborhood (working-class) scale while insisting on the Americanism of the city as a whole. The scripting of the revanchist city, however, is viscerally local and global, no longer so isolated or insulated, if indeed it ever was.

AFTER TOMPKINS SQUARE PARK: NEW YORK'S HOMELESS WARS

A real estate market on fire during the 1980s, New York's Upper West Side, like the Lower East Side, experienced a significant retrenchment of gentrification by 1989. Indeed it was in this neighborhood that the idea of "de-gentrification" seems to have been coined. The slowing of gentrification eased the rate of eviction and the pace of rent increases, and although there is disagreement on actual numbers, most commentators suggest that the home-less population for New York City as a whole has stabilized since the beginning of the 1990s. But by the same token, the liberal concern for homeless people, kindled initially by the surge in homelessness in the 1980s expansion, and nurtured in neighborhoods like the Upper West Side, began to diffuse.

"We're not suicidal liberals anymore," exhorts a neighborhood activist in a front-cover *New York* story about "the decline and fall of the Upper West Side." A traditionally liberal counterpart to the stuffier *New Yorker*, *New York* magazine has championed various liberal causes, and finds a natural clientele in the Upper West Side, which has experienced various waves of gentrification since the 1960s. But *New York* is clearly fed up with homelessness. Bemoaning the cooling of gentrification and the increased bankruptcy of small businesses, writer Jeffrey Goldberg proposes that the last few years had seen an influx of homeless people to the neighborhood, drawn by the availability of social services: "small business is no longer the dominant industry in the Upper West Side. Homelessness is" (Goldberg 1994: 38). Anxious not to appear racist, he quickly cites the "environmental racism" of city policies that locate the preponderance of homeless and other social service facilities in poor neighborhoods, while effectively arguing that such facilities should not be pro-vided at all. And he finds a Republican "community advocate," prominently identified as black, to support his case:

> The brain-dead liberal attitude is to look at the homeless and say, "We must house them here." And then you wake up the next morning and there are more homeless on the street, because the more we take the more the city sends. Soon this neighborhood will be made up entirely of social service sites and people in expensive co-ops.
>
> (Goldberg 1994: 39)

But if New York's "homeless wars" have a geographical focus it is surely the Lower East Side – more specifically, Tompkins Square Park. And it has to be

Plate 10.1 After the park (*The Shadow*)

said that the clearance of Tompkins Square Park was presided over not by the more reactionary mayors in the city's recent history but by David Dinkins. Dinkins, a liberal Democrat and sometime member of Democratic Socialists of America, was elected with strong support from New York City's housing and antihomelessness movement, but quickly sanctioned the first evictions of homeless people from the park in December 1989 – weeks after his election, but before his inauguration. This initiated a four-year corrosion of Dinkins' connections with the mass support that had elected him. As the *Village Voice* noted of the 1991 evictions, for "the homeless residents, many of them now scattered in abandoned lots around the park, the closing of the park was just one more betrayal for an administration they thought would stand up for the rights of the poor" (Ferguson 1991a: 16). In finally closing the park, Dinkins borrowed a script not from housing or homeless advocates but from the editorial pages of the *New York Times*, which quoted the Webster's dictionary definition of "park" then judged that Tompkins Square was no park at all: "A park is not a shantytown. It is not a campground, a homeless shelter, a shooting gallery for drug addicts or a political problem. Unless it is Tompkins Square Park in Manhattan's East Village." Homeless residents of the park, according to the *Times*, had "stolen it from the public" and the park would have to be "reclaimed." Just three days before the park's closure, the newspaper inveighed against further partial solutions, preferring instead a "clean sweep" as "the wiser course though riskier politically." There were, it seems, "some legitimately homeless people" who "live in the park," and therefore "misplaced sympathy abounds" ("Make Tompkins Square . . . " 1991). In an interview for National Public Radio, Parks Commissioner Betsy Gotbaum borrowed from the same script, adding her own racial coding of the new urban frontier: "It was filled with tents, even a tepee at one point. . . . It was really disgusting."

The closing of Tompkins Square Park on June 3, 1991, and the eviction of more than 300 homeless residents fanned the issues of homelessness throughout the Lower East Side, and the locus of political action likewise spread out from the park as the entire neighborhood became the contested zone and the neighboring streets became a shifting DMZ (demilitarized zone). In the summer of 1991, in the immediate surrounds of the fenced park, a nightly ritual of "walk the pig" ensued. It is worth quoting at length an eyewitness report of just one incident, which offers a visceral portrait of the agency behind the revanchist city:

> Since the police takeover June 3, there have been nightly gatherings on the steps of St. Brigids Church on Avenue B [on the southeast side of the park], a focal point of community resistance. On Friday, a dozen parents with their children gathered among the punks and anarchists and tried to march against the line of riot police blocking their way, chanting "Open the park!" When they were forced back onto the sidewalks, some 800 residents took to the streets, banging on drums and garbage can lids, and leading the police cordon [protecting the park] that dutifully followed them from Loisaida through the West Village and back through

the projects off Avenue D – what locals call the nightly "walk the pig" routine.

They were confronted on the steps of St. Brigids by at least 100 cops, who beamed blinding high-intensity lights into the crowd. The protesters remained peaceful until two undercover cops shoved their way into the church entrance on Avenue B, claiming they wanted to inspect the roof for bottle throwers. One parishioner, Maria Tornin, was struck in the face and knocked against the stairs by one of the cops, and Father Pat Maloney of Lazarus Community was shoved against the wall. Backed by his parishioners, St. Brigids' Father Kuhn pushed the undercover cops out the door.

"When the law ends, tyranny begins, and these guys are tyrants," shouted Father Maloney, leading an angry mob to the paddy wagon where the undercovers had fled. . . .

Last Saturday, as bulldozers rumbled past the ripped-up benches and shattered chess tables [in the cordoned-off park], a second demonstration of over 1000 Lower East Side residents linked arms around the park. As the church bells of St. Brigids rang out, dreadlocked anarchists in combat boots and nose rings held hands with Jewish grandmothers in print dresses and plastic pearls in a peaceful show of unity not seen since the 1988 police riot.

(Ferguson 1991b: 25)

The closure of Tompkins Square Park in June 1991 marked the onset of a stern antihomeless and antisquatter policy throughout the city that readily expressed the ethos of the revanchist city. Spearheaded by "Operation Restore" in the Lower East Side, the new antihomeless policy initiated by the City administration in 1991 was intended to "take back" the parks, streets and neighborhoods from those who had supposedly "stolen" them from "the public." The attack on Lower East Side squatters had been escalated first in 1989, but two years later, with perhaps 500 to 700 squatters still in thirty to forty buildings in the Lower East Side at the beginning of 1992, it actually proved too difficult for the City to evict the squatters although several buildings in the neighborhood as well as in the Bronx were cleared. Attention instead was focused on a major campaign described by the *New York Times* as a "crackdown on homeless" (Roberts 1991a).

Homeless people had responded to the park closure by immediately reerecting shanties and tent cities on several empty lots in the neighborhood, generally in the poorer, still largely Puerto Rican neighborhood to the east of the park. As it grew, this "Dinkinsville," as it was dubbed, was subjected to surveillance and eventually bulldozing, beginning in October 1991 with a "sweep" of three lots, and the reeviction of 200 people (Morgan 1991). Like the park itself, these sites were quickly fenced to prevent public squatting by homeless people on empty space. Once again the evictees were moved further east, setting up or joining encampments under the Brooklyn, Manhattan and Williamsburg Bridges, under the FDR Drive, or in any available space defensible from public view, police attack and bad weather. Fire destroyed the encampment under one East River Bridge, killing one resident, and a year

Plate 10.2 Bulldozing Dinkinsville. Third and Fourth Streets between Avenues C and D, New York (John Penley)

later, in August 1993, the City bulldozed "the Hill" below another – a well-established shantytown of fifty to seventy residents described as "one of the most visible symbols of homelessness in Manhattan" (Fisher 1993). Squeezed yet further east again, many evictees scattered up and down the waterfront of the East River and into Sara Delano Roosevelt Park, only to be moved yet again in 1994 when a supposed reconstruction of part of the waterfront began. The dispersal of homeless people from the Lower East Side to temporary and fragmented sites throughout Manhattan and beyond was essentially complete by 1994.

Elsewhere in the city, attacks on homeless people also gathered momentum. Shantytowns under the West Side Highway, at Columbus Circle and in Penn Station were simultaneously razed beginning in the autumn of 1991. And the "Mole People" were discovered. The police sweeps and fires led to the revelation in the local press of whole vistas of homeless life previously "unknown," including several encampments under bridges and in transportation and utility tunnels, underground. In some cases these were long-term encampments whose residents were treated in press reports as simultaneously exotic and dehumanized. The longer they had lived "underground" the more attractive they were to journalists, especially if they also held down steady jobs. The label that these homeless people sardonically applied to themselves – the Mole People – was reapplied with brute empirical descriptiveness by the press (see Toth 1993).

To match emerging hardline policies concerning outdoor public space, the Transit Authority instituted new antihomeless policies for its major hubs, aimed at beginning to deny homeless people access to indoor public space. At Grand Central Station a more novel approach was tried. Formed in the wake of Mobil Oil's departure from Manhattan and their parting movie, which depicted the tribulations of a white suburban executive trying to commute to work through mobs of harassing homeless people, the "Grand Central Partnership" was formed to deal with "the problem." Funded by levies from local businesses, the Partnership began with private security patrols, and cajoled homeless people out of the area by offering food and shelter in a nearby church. This model of "Business Improvement Districts" was replicated throughout the metropolitan area. In the meantime, the Grand Central Partnership has participated in "sweeps" of homeless people and shanties from First Avenue, and is under investigation for allegedly employing some evictees to forcibly remove others.

Eviction, in fact, represented the only true homeless policy of the Dinkins administration between 1990 and 1993; it was, more appropriately, an antihomeless policy. As the "crackdown" began in late 1991, the director of the mayor's Office on Homelessness became increasingly frustrated. Well-meaning in the beginning but finding herself increasingly drawn into the administration's only real strategy – blaming homeless people for their lack of homes – she resigned. By 1993, with hundreds of homeless people sleeping overnight in City offices, the City administration and several bureaucrats were found in contempt of court for their lack of a homeless policy and failure to provide court-ordered shelter.

The bankruptcy of liberal homeless policy had become obvious at the national level during the 1980s when there was little effective response to Reaganism. By the end of that decade, the failure of liberal urban policy was being played out at the local level. This failure is not explicable in simple financial or technical terms. Whatever very real financial constraints were imposed – by Reagan/Bush attacks on social spending, by the inability of a shelter system with a capacity of approximately 24,000 people to house as many as four times that number, or indeed by the sheer inability of any city bureaucracy to make the shelters safe places – the failure of liberalism is primarily a failure of political will, and it infected neighborhoods dealing with large numbers of homeless people as much as the city administration. The failure of liberal urban policy also systematically disrupted the systems of social reproduction (Susser 1993).

In the Lower East Side, community support for those living in Tompkins Square Park clearly eroded as the encampment became more entrenched and the City provided no solutions. The park was workplace and playspace, living room and bathroom, for hundreds of people daily, and the result was hardly a salubrious solution to emergency housing and other social needs. Even a sympathetic observer had to conclude when the park was closed that

> the situation had reached a crisis point that even the tolerant Lower East Side milieu could no longer sustain. . . . Most residents are too fed up with the homeless and the park to put up another fight. And the community surrounding the park has already changed.
>
> (Ferguson 1991a)

What the Dinkins administration shared with frustrated residents of the Upper West Side and the Lower East Side – residents whose self-interest rapidly exfoliated in response to the continued presence of homeless people – was the assumption that homelessness is an unfortunate occurrence which, however systemic, represents a moral wrong and can be resolved with *ad hoc* homeless policies. The transparency of that assumption can be maintained as long as homeless policies exact little or no cost on the housed; when homelessness begins to hurt the housed in any significant way, the political will to house everyone has to be built on a stronger political and analytical foundation than moral sympathy.

The erosion of sympathetic support and action – both in the city administration and among city dwellers – came in the context of a significant economic depression which threatened many people's livelihoods and identities, whether they were housed or unhoused. Sympathy for "the homeless" – a nomenclature that objectified, distanced and inured, all in one – became a luxury that fewer and fewer people would allow themselves. It was on this foundation of the abject failure of liberal urban policy to deal with homelessness that a newly elected Mayor Rudy Giuliani set about consolidating the emerging revanchist city in 1994. The city's first Republican mayor in a quarter-century, Giuliani began his mayoralty with a concerted attack on homeless people. The opening volley came days after taking office when, according to a widespread story, Giuliani was questioned by a reporter about the frigid cold spell that had descended on the city. Given the number of homeless people on the streets, he was asked, what did the mayor intend to do? "We're working on the weather," the mayor reportedly responded.

But the Giuliani administration was in reality working on a lot more than the weather. He immediately announced plans to outlaw squeegee windshield cleaning by homeless people as well as panhandling around the city, and initiated a demeaning subway poster campaign aimed at humiliating homeless beggars and intimidating other passengers: "Don't give them your money," blared the posters, over images of homeless people as alternatively debased or threatening. Giuliani's first budget proposal included a provision for homeless people to pay "rent" for their nights in City-run shelters, and for the barring of people from shelters if they refused medical, drug and alcohol rehabilitation, and other social service referrals.

In the weeks after Giuliani's election, amid escalated police "sweeps" of public spaces, staff at the nonprofit advocacy group Coalition for the Homeless began noticing a pattern of increased injuries among homeless clients, and formed an organization called Streetwatch which both monitored police treatment of homeless people and gathered testimony from homeless people. After several months, Streetwatch filed a multi-million-dollar suit alleging intensified police harassment, abuse and brutality in Penn Station under the Giuliani administration, documenting some fifty allegations. "The complaint reads like pages torn from *A Clockwork Orange*," noted the *Village Voice* (Kaplan 1994a). One complaint gathered by Streetwatch includes the following testimony:

"I was seated in the waiting area [at Penn Station]. . . . Two white male officers . . . approached me. They said 'get up and leave right now or we're going to help you leave. . . . ' I bent down to pick [my bag] up. They grabbed me by the elbows and flung me against the concrete pillar next to the chairs. They knocked out my front tooth, and opened up a large gash over my right eyebrow, which was bleeding profusely. It also dislocated my nose, and broke my eyeglasses. . . . They told me that they didn't want to see me in there anymore, and if they did I would be 'crawling for quite a while'. Then they . . . pushed me out the door violently so that I fell to the sidewalk and hit the back of my head. That injury required eight stitches. . . . Ever since the fall, I've had dizzy spells and blackouts."

(quoted in Kaplan 1994b)

The police campaign to remove homeless people from public places was largely justified in terms of a Giuliani initiative to criminalize a broad swath of activities as inimical to "the quality of life" in city neighborhoods. The rubric of "quality of life" offenses has given the New York City Police Department unprecedented powers to remove homeless people from certain streets, install them in unsafe shelters, or simply force them into hiding. The administration has also cut funding for soup kitchens for homeless people. With the move against several buildings on Thirteenth Street in May 1995, the City has again initiated a concerted effort to clear squatters out of otherwise vacant Lower East Side buildings.

The erosion of support for homeless people in New York in the early 1990s also came amid a broader discovery in the *national* media that in liberal as much as not-so-liberal neighborhoods, "a growing national ambivalence about the homeless" had become pervasive (Roberts 1991b). Beginning in more conservative cities from Miami to Atlanta, but quickly adopted in more liberal bastions such as Seattle and San Francisco, harsh measures against sleeping and camping in public pavement sitting, panhandling and wind shield washing have been enacted (Egan 1993). And in an effort "to make downtown Los Angeles friendlier to business, the city administration is working on a plan to shuttle homeless people to an urban campground on a fenced lot in an industrial area" ("Los Angeles plans a camp . . . " 1994). The revanchist city stretches well beyond New York, and the criminalization of more and more aspects of the everyday life of homeless people is increasingly pervasive. The US press, in the meantime, has run out of new angles on the visceral realities of homelessness, and newspapers have either continued to run increasingly anemic, predictable stories of the streets or else eschewed the issue altogether.

A certain neoliberal revisionism also set in among an array of academic commentators. In the absence of any significant housing initiatives for resolving homelessness – locally as well as nationally – discussion of causes increasingly reverted to aspects of individual behavior rather than societal shifts, and the practice of blaming the victims quietly gained credence in erstwhile liberal circles (cf. Rossi 1989). Attempts were made to deny that homelessness was any kind of growing problem at all in the 1980s (White

1991), with the clear implication that existing policies, predicated on a sense of emergent crisis of homelessness in the 1980s, are wrongheaded. In his recent book *The Homeless* (Jencks 1994c) and in a pair of articles in the traditionally liberal *New York Review of Books* (Jencks 1994a, 1994b), Christopher Jencks distances himself only partially from this revisionism. Defending much lower numerical estimates of homelessness – 324,000 nationally in 1990 compared with widespread estimates ranging up to ten times this figure – Jencks cautions that the visual evidence of greatly increased homelessness in the streets may be misleading:

> But what we see on the streets often depends more on police practices than on the frequency of destitution. The number of panhandlers, for example, depends mainly on the risk of arrest and how much one can earn from panhandling compared to other activities. Most panhandlers appear to live in conventional housing, and only a minority of the homeless admit to panhandling. Nor is appearance a reliable indicator of homelessness. Rossi's interviewers rated more than half their respondents "neat and clean."
>
> (Jencks 1994a: 22)

Even the "numbers game" of estimating the extent of homeless people comes to be dominated by middle-class stereotypes about "appearance," the morality of begging, and discredited assumptions of rational behavior from economic choice theory.

That his analysis should rely on a theory of economic rationality is ironic given Jencks' acknowledgment that "No other affluent country has abandoned its mentally ill" (1994a: 24) to the extent that the US has, and his conclusion that mental illness plays a large part in homelessness. In fact, Jencks identifies several behavioral and structural factors contributing to any increases in homelessness in the 1980s: declining marriage rates for young mothers; the crack epidemic; deinstitutionalization of the mentally ill; declining demand for unskilled workers; personal preference for the streets; and various changes in local housing and shelter legislation that "encouraged" people to become homeless while preventing the private market from supplying a clear need (Jencks 1994a, 1994b, 1994c). Whatever the conservatism of this argument, Jencks retains a sense of liberal responsibility for housing homeless people, and he indicts the present shelter system for, above anything else, prohibiting privacy. He suggests with steely pragmatism that the only realistic solution "is to build cubicle hotels" (1994b: 44). Appealing to the model of Chicago in the 1950s – "windowless five- by seven-foot rooms furnished with a bed, a chair, and a bare light bulb" (1994b: 39) – Jencks argues for the rolling back of housing codes and if necessary the subsidy of entrepreneurs to build such cubicle hotels. Social services would be provided through a voucher system in exchange for work. The model here again is "rational choice":

> In 1958, a cubicle cost less than a six-pack of beer, making privacy cheaper than oblivion. By 1992 a six-pack cost less than what a cubicle cost, making oblivion cheaper than privacy. The same pattern obtains if we compare the price of a cubicle to the price of cocaine.
>
> (Jencks 1994b: 39)

Jencks' vision may qualify as the archetypal neoliberal revanchism *vis-à-vis* homeless people: a residue of sympathy activated by thinly disguised hatred and abhorrence.

The reaction against homelessness and homeless people in the 1990s represents only one aspect of the emerging revanchist city, if a particularly nasty one. It is not that political support for homeless people has entirely vanished nor that more critical treatments of homelessness are no longer written (see for example Hoch and Slayton 1989; Wagner 1993). Rather, the dominant discourse on homelessness has moved decisively away from the sympathetic albeit often patronizing stance of the late 1980s to a more brazen indictment of homeless people not just for their own predicament but for larger social ills. In this classically revengeful conservatism, the connections between societal process and individual predicament are reversed.

The rallying cry of the revanchist city might well be: "Who lost the city? And on whom is revenge to be exacted?" Expressed in the physical, legal and rhetorical campaigns against scapegoats, identified in terms of class, race, gender, nationality, sexual preference, this reaction scripts everyday life, political administration and media representations of the contemporary US city with increasing intensity. The revanchist city is, to be sure, a dual and divided city of wealth and poverty (Mollenkopf and Castells 1991; Fainstein *et al.* 1992), and it will continue to be so as seemingly apocalyptic visions of urban fissure, anticipated by Davis (1991) and realized in the Los Angeles uprising, appear more and more realistic. But it is more. It is a divided city where the victors are increasingly defensive of their privilege, such as it is, and increasingly vicious defending it. The revanchist city is more than the dual city, in race and class terms. The benign neglect of "the other half," so dominant in the liberal rhetoric of the 1950s and 1960s, has been superseded by a more active viciousness that attempts to criminalize a whole range of "behavior," individually defined, and to blame the failure of post-1968 urban policy on the populations it was supposed to assist.

DEGENTRIFICATION?

There can be little doubt about the depth of the crisis that hit urban real estate after 1989. Fewer new housing units were built in New York City in 1991 (7,639) than in any year since World War II. Prices plummeted and even some rents dropped; housing foreclosures increased (Ravo 1992a, 1992b); and disinvestment made a strong resurgence with tax arrears increasing by an estimated 71 percent between 1988 and 1992 (Community Service Society 1993). In the Lower East Side the number of buildings five or more quarters in arrears almost tripled between 1988 and 1993, and a number of smaller land-lords were forced to sell out to what activists have called "the bottom feeders" of the real estate market (Benjamin Dulchin, personal communication, 30 September 1994). These larger investors make a specialty of buying out smaller landlords strapped for cash. These were the headlines that gave rise to the fears of degentrification. While they were cause for major handwring-ing among real estate professionals, at least one commentator was publicly relieved. Peter Marcuse (1991) argued that "the death of gentrification may be

Plate 10.3 The revanchist city: police attack demonstrators at the Community Board 3 meeting, New York, June 22, 1993 (*The Shadow*)

just what the doctor ordered." The "receding tide of gentrification may leave stranded some folks who, out of speculative greed or an honest search for good housing, rushed in where the market has (so far) feared to tread." And yet here is a golden opportunity, Marcuse concluded, for the City to pick up properties cheaply and turn them over to much-needed low-income housing. He even proposed that the infamous Christodora on Tompkins Square Park, where prices had fallen dramatically, be rehabilitated for just such a purpose.

Unfortunately, neither Marcuse's optimism nor the real estate industry's pessimism is warranted. In the first place, the collapse of the real estate market between 1989 and 1993 was very uneven. Of all the new houses constructed in 1991, most were in Manhattan, suggesting a continued recentralization of real estate activity, albeit at a smaller scale, consistent with the continuation of gentrification. Conversely, it seems that an unprecedented share of foreclosures were in the suburbs (Ravo 1992b). Second, it was the properties at the top of the market where speculation had been most extreme that suffered the greatest price and rent declines. If a 30–40 per cent price drop was not unusual at the top end in this period, it was not matched by significant rent increases at the bottom of the market. The shortage of low-income housing was ameliorated and prices stabilized, but cheaper housing in the early 1990s was subject to very different dynamics. The depression at best provided a modicum of breathing space for poorer tenants. Evictions in the Lower East Side, for example, were down significantly as landlords, no longer in a renter's market, preferred to wait out the real estate crisis with full apartments and take what they could get.

This bout of so-called "degentrification" will in all likelihood register as a passing lapse in an otherwise fervid rebuilding of certain districts of the central city such that a more segregated geography and a revanchist politics of the city unfold together. Thus by 1994 the rental market had a resurgence, especially in the Upper West Side, where landlords were not only pushing up rents but again actively trying to evict tenants from rent-controlled and stabilized apartments. In the Lower East Side the dramatic increase in foreclosures had flattened out by late 1993, and if tax arrears continued to increase, this mostly reflected the City's now publicly declared policy of no longer taking buildings *in rem*.

On the cultural front too preparations were being made for a regentrification. As the rhetoric of "degentrification" was quickly erasing the violence of gentrification, the treatment of regentrification in the 1990s intensified the erasure. Thus the reopening of Tompkins Square Park in 1992 was accompanied by a predictable naturalization of Tompkins Square's history, geography and culture in the press. Noting the parallels with Central Park in the 1930s (read: forget 1988 to 1991), the *New York Times* immediately heralded the reconstructed Tompkins Square as a "shining emerald" (Bennet 1992). Within a year, in fact, the estheticization of the neighborhood was in full swing with photographs in the press of young, white middle-class families enjoying the park once again. An article in the fashion pages of the *Times* celebrated the neighborhood as a "serendipitous, ad hoc mall" in the inner city, noting, without any mention at all of the preceding conflicts, that the rehabilitation of the park was followed by a thoroughgoing "fashion rehab." Despite the lingering depression affecting the region, more than twenty-five new shops opened in the year after the park's reopening, readying themselves again "for the neighborhood's inevitable onset of young professionals." "Since the renovation of Tompkins Square Park . . . the area . . . has become extremely desirable from a commercial point of view. . . . With rents from $20 to $30 a square foot (and climbing), there is sudden interest 'from successful businesses in the West Village and SoHo who want to relocate,'" observes one local broker (quoted in Servin 1993).

In this context, the argument for "degentrification" seems at best premature (see also Badcock 1993, 1995; Lees and Bondi 1995). Predictions of the demise of gentrification are premised on essentially consumption-side explanations of the process, in which any pickup in the economic demand is magically converted into a long-term trend. If, however, the patterns of capital investment and disinvestment are at least as important in creating the opportunity and possibility for gentrification, then a rather different vision emerges. The decline in housing and land prices since 1989 has been accompanied by a disinvestment from older housing stock – repairs and maintenance unperformed, building abandonment – and these are precisely the conditions which led in the first place to the availability of a comparatively cheap housing stock in central locations. Far from ending gentrification, the depression of the late 1980s and early 1990s may well enhance the possibilities for reinvestment. Whether gentrification resurges following the economic depression now appears to be a significant test of production-side versus consumption-side theories. At the very least, the deflation of gentrification activity following the 1987 stock market crash should be seen as a dramatic reassertion of economics in the land and housing markets.

The language of degentrification, of course, not only justifies the political momentum behind the revanchist city, but feeds the self-interest of real estate developers and contractors. "Gentrification" has indeed become a "dirty word." It expresses well the class dimensions of recent inner-urban change, and it is hardly surprising that real estate professionals took advantage of a very real slowdown in gentrification to attempt to expunge the word and the memory of the word's politics from the popular discourse. But neither the memory nor the profits of gentrification are likely to be erased so quickly. Indeed, it may not be too much of an exaggeration to surmise that proclaiming the end of gentrification today may be akin to anticipating the end of suburbanization in 1933.

The resurgence of gentrification following the economic depression of 1989–1993, and the expansion of homelessness to which it contributes, will not mean the end of the revanchist city and the instigation of a kinder, gentler urbanism. The more likely scenario is of a sharpened bipolarity of the city in which white middle-class assumptions about civil society retrench as a narrow set of social norms against which everyone else is found dangerously wanting; and, by way of corollary, we can expect a deepening villainization of working-class, minority, homeless and many immigrant residents of the city, through interlocking scripts of violence, drugs and crime.

Thus in the spring of 1995, faced by a $3.1 billion budget deficit, Mayor Rudy Giuliani voiced explicitly a long-implicit intent of service and budget cuts. By cutting services, the mayor told a small group of newspaper editors, he hoped to encourage the poorest of the city's population, those most dependent on public services, to move out of the city. Shrinkage of the poor population would be a "good thing" for the city, he suggested. "That's not an unspoken part of our strategy," he added. "That is our strategy" (quoted in Barrett 1995).

RETAKING THE URBAN FRONTIER

"Extermination," George Custer declared in 1865, eleven years before his last stand in the Dakotas, "is the only true policy we can adopt toward the political leaders of the [Sioux] rebellion." "Then, and not till then," he concluded, "may the avenging angel sheathe his sword, and our country will emerge from this struggle regenerated" (quoted in Slotkin 1985: 384). The "regeneration" of the *fin-de-millénaire* city is premised on a similar agenda of extermination. There are now too many reports of homeless people being set upon or set alight to record them as freak attacks by rogue citizens. If Custer's brand of extermination is now less than polite, even in the revanchist city, homeless people suffer a symbolic extermination and erasure that may leave them alive but struggling on a daily basis to create a life with any quality at all. Neither gentrification nor degentrification solves *their* problems.

As Peter Marcuse (1991) has put it, "The opposite of gentrification should

Plate 10.4 Demonstration against eviction, Park Avenue, New York
December 22, 1991 (*The Shadow*)

not be decay and abandonment" – degentrification – "but the democratization of housing." And the democratization of housing will hardly come through the kind of amelioration of development represented by "double-cross subsidy" programs, "proactive neighborhood development," or the election of liberal urban regimes – like that of Dinkins. Their power to make real change has diffused entirely in the revanchist city. Liberalism is without portfolio. The city has become a "bubbling cauldron" (M. Smith and Feagùr 1996), of which anti-gentrification resistance is a part (McGee 1991). In a perverse way, however, the frontier mythology supports more direct alternatives, whether in Tompkins Square Park or Hamburg's Hafenstrasse, where a militant squatters' commune resisted "renovation, fascism and the police state," to quote one of their slogans, into the mid-1990s.

For whatever the powerful patriotic rhetoric of national regeneration, there are two sides to every frontier. Otherwise it would not be a frontier – economic, cultural, political, geographical. And in the case of Custer versus the Sioux, I would bet that by the end of the twentieth century, most of us would have to come down on the side of the Sioux – our childhood indoctrination by Hollywood notwithstanding. Custer's extermination speech came only three years after the Homesteading Act, which granted rights to land to western "pioneers." This Act was hardly a beneficent deed by a beloved government. Dressed in the best rhetoric of pioneering and rugged individualism, it was the best compromise the government could exact in a losing situation. Prior to 1862, the majority of heroic pioneers were actually illegal squatters who were democratizing land for themselves. They took the land they needed to make a living, and they organized clubs to defend their land claims against speculators and land grabbers, established basic welfare circles, and encouraged other squatters to settle because strength lay in numbers. Squatters' organization was the key to their political power, and it was in the face of this organization and rampant squatting on the frontier that the Homesteading Act of 1862 was passed.

The whole force of the myth has been to dull this class inscription of the frontier, erase the central threat to authority that the frontier posed, by swaddling it in the romantic cloak of individualism and patriotism. If we are truly to embrace the city as the new frontier today, then the first and most patriotic act in pioneering, if historical accuracy is to be observed, will be squatting. It is just possible that in a future world we may also come to recognize today's squatters as the ones with a more enlightened vision about the urban frontier. That the city is become a new Wild West may be regrettable, but it is surely beyond dispute; what kind of Wild West is precisely what is being fought out.

NOTES

CHAPTER 1 "CLASS STRUGGLE ON AVENUE B" (pp. 3–29)

1 The poet Allen Ginsberg relates this reaction from a visiting Chinese student who had been in Tiananmen Square during the student confrontations with police that June. In China the police "were dressed in cloth like everyone else. He said the contrast was *amazing*, because in China it was pushing back and forth, and maybe batons. But here it was people who looked like they were dropped from outer space with these helmets on, dropped in the middle of the street from outer space and just beating people up, passers-by and householders – anyone in their path. Completely alienated and complete aliens" ('A Talk with Allen Ginsberg" 1988; emphasis in original).

2 Despite his critique, Owens (1984: 163) cops out at the end: "Artists are not, of course, responsible for 'gentrification'; they are often its victims." As Deutsche and Ryan comment: "To portray artists as the victims of gentrification is to mock the plight of the neighborhood's real victims" (1984: 104).

3 Zeckendorf is the scion of a major real estate family that has long been involved in gentrification. His father, William Zeckendorf Sr., was the major developer behind Society Hill Towers, Philadelphia (see pp. 127–128).

4 The words quoted come from Moufarrege (1982).

CHAPTER 2 IS GENTRIFICATION A DIRTY WORD? (pp. 30–47)

1 This literature is too vast to replicate fully here. For the most recent round of debates, see Hamnett (1991), Bondi (1991a, 1991b), Smith (1992, 1995c), van Weesep (1994), Lees (1994), Clark (1994), Boyle (1995d). These works mostly cover the pivotal references in the fifteen-year debate.

2 Consider the following account: "A professor from South Korea, who is head of the writers' union there, got up and said that, well, in order to understand what he is going to talk about, he had to talk a little about himself. 'Normally I feel terribly embarrassed about doing that sort of thing,' he said. 'But in the States I have found a very easy way to do it; it's called revealing your subject positionality'" (Haraway and Harvey 1995).

CHAPTER 3 LOCAL ARGUMENTS (pp. 51–74)

1 "Suburbs" here implies the area outside the city boundary but inside the SMSA as it was defined at the time. The older suburbs that now appear inside the city as a result of subsequent annexations are therefore counted as sections of the city. This definition is justified here since one of the main selling points of gentrification is

that it will bring additional tax revenues to the city. Clearly, annexed suburbs already pay their taxes to the city.

2 I omit speculators here for the obvious reason that they invest no productive capital. They simply buy property in the hope of selling it at a higher price to developers. Speculators do not produce any transformation in the urban structure.

3 I am grateful to Bob Beauregard, who brought this piece to my attention in the course of his own research for *Voices of Decline* (Beauregard 1993).

CHAPTER 6 MARKET, STATE AND IDEOLOGY (pp. 119–139)

1 The objectives of the renewal plan are made explicit in: Redevelopment Authority of Philadelphia, Final Project Reports for Washington Square East Urban Renewal Area, Units 1, 2 and 3. Unpublished, undated. Like much of the empirical research for this chapter, this document was researched in the RAP's files on the "Washington Square East Urban Renewal Area," as the Society Hill Project was officially called.

2 The Old Philadelphia Development Corporation employed a Madison Avenue professional to advertise Society Hill nationally (OPDC Annual Report 1970).

3 Another Redevelopment Authority executive director, Augustine Salvitti, was indicted on sixteen charges for allegedly receiving a $27,500 kickback in return for a $575,000 Redevelopment Authority contract. Salvitti was charged with "one count of racketeering, one count of extortion and 14 counts of mail fraud." *Philadelphia Inquirer*, October 29, 1977.

4 Redevelopment Authority Monthly Project Balance for Washington Square East, July 31, 1976.

5 I. M. Pei and Associates, "Society Hill, Philadelphia: a plan for redevelopment," prepared for Webb and Knapp, undated.

CHAPTER 7 CATCH-22 (pp. 140–164)

1 Comments by Harold Wallace as he gave a guided walk round parts of central Harlem in 1989. Quoted in Monique Michelle Taylor, "Home to Harlem: black identity and the gentrification of Harlem." Ph.D. dissertation, Department of Sociology, Harvard University, 1991, p. 7.

2 Chall actually underestimates the extent of gentrification. He makes much of aggregate citywide data which do not show any absolute reversal of suburbanization trends (at least up to 1980), and so downplays the extent of gentrification. In fact, if one examines per capita income changes, rather than household income, and if one is prepared to examine spatially contiguous groups of census tracts and their internal changes, a much clearer picture of gentrification emerges. The most significant aspect of the 1980 census in this respect is precisely that gentrification shows up at the census tract level for the first time. This is immediately evident by mapping per capita income and median contract rent increases for Manhattan, where the southern and western part of the island show dramatic rises. The findings demonstrate the substantial spread and expansion of the process since 1970. See also Marcuse (1986).

3 Interview with Roy Miller, the then director of the Harlem Office of Community Development and Neighborhood Preservation, April 13, 1984.

4 Interview with Donald Cogsville, president of the Harlem Urban Development Corporation, April 20, 1985.

5 Interview with Donald Cogsville, April 20, 1985.

CHAPTER 8 ON GENERALITIES AND EXCEPTIONS (pp. 165–186)

1 Although condominium conversion in the US and tenure conversion from private rental to private ownership in Britain are broadly if not always connected to gentrification, in the Netherlands it is important to be more circumspect about this connection. Given the strict regulation of the rental market in prior years, condominium conversion in the early 1980s became a means of escape – actually a means of disinvestment – as much as a gentrification strategy (van Weesep and Maas 1984: 1153). It is not that condominium conversion has not been a vehicle for owner disinvestment in the US or UK – the post-1987 experience has exposed many such cases – but that the Netherlands represents a paradoxical extreme. In fact, since condominium and housing prices continued to increase rapidly until between 1987 and 1989 in New York and London, and to a lesser extent in Amsterdam, the gains and the gentrification resulting from condominium and tenure conversion largely outpaced disinvestment, which did, nonetheless, reassert itself somewhat in the first half of the 1990s.

2 About one-third of the housing stock in this eastern part of the Old City was built after 1980 (van Weesep and Wiegersma 1991: 102).

3 For a similar argument *vis-à-vis* New York's Lower East Side, see Lees (1989).

CHAPTER 9 MAPPING THE GENTRIFICATION FRONTIER (pp. 189–209)

1 The Department of City Planning of the City of New York has compiled a centralized computerized database, MISLAND, covering a wide range of information about the city. The information is provided in a series of separate files. The arrears data are taken from the Property Transaction and Real Property Files and the vacancy data from the Con Edison File. For the latter, the figures include three categories of account: current accounts; accounts awaiting turn-on; and accounts in vacant buildings. The latter are included among active accounts; Con Edison defines multiunit buildings as vacant if their vacancy rate is over 40 percent.

2 The mapping was accomplished using the Surfer trend surface analysis package produced by Golden Software.

BIBLIOGRAPHY

"A talk with Allen Ginsberg" (1988) *The New Common Good*, September.

Abrams, C. (1965) *The City Is the Frontier*, New York: Harper and Row.

Adde, L. (1969) *Nine Cities: Anatomy of Downtown Renewal*, Washington, DC: Urban Land Institute.

Advisory Council on Historic Preservation (1980) *Report to the President and the Congress of the United States*, Washington, DC: Government Printing Office.

"After eviction, Paris homeless battle police" (1993) *New York Times* December 14.

Aglietta, M. (1979) *A Theory of Capitalist Regulation*, London: New Left Review.

AKRF, Inc. (1982) "Harlem area redevelopment study: gentrification in Harlem," prepared for Harlem Urban Development Corporation.

Allen, I. L. (1984) "The ideology of dense neighborhood redevelopment," in J. J. Palen and B. London (eds) *Gentrification, Displacement and Neighborhood Revitalization*, Albany: State University of New York Press.

Allison, J. (1995) "Rethinking gentrification: looking beyond the chaotic concept," unpublished Ph.D. dissertation, Queensland University of Technology, Brisbane.

Alonso, W. (1960) "A theory of the urban land market," *Papers and Proceedings of the Regional Science Association* 6: 149–157.

—— (1964) *Location and Land Use*, Cambridge, Mass.: Harvard University Press.

Anderson, J. (1982) *This Was Harlem: A Cultural Portrait. 1900–1950*, New York: Farrar Straus Giroux.

Anderson, M. (1964) *The Federal Bulldozer: A Critical Analysis of Urban Renewal, 1949–1962*, Cambridge, Mass.: MIT Press.

Aparecida de Souza, M. A. (1994) *A Identidade da Metrópole*. São Paulo: Editora da Universidade de São Paulo.

Aronowitz, S. (1979) "The professional-managerial class or middle strata," in P. Walker (ed.) *Between Labor and Capital*, Boston: South End Press.

Bach, V. and West, S. Y. (1993) "Housing on the block: disinvestment and abandonment risks in New York City neighborhoods," New York: Community Service Society of New York.

Badcock, B. (1989) "Smith's rent gap hypothesis: an Australian view," *Annals of the Association of American Geographers* 79: 125–145.

—— (1990) "On the non-existence of the rent gap: a reply," *Annals of the Association of American Geographers* 80: 459–461.

—— (1992a) "Adelaide's heart transplant, 1970–88: 1 creation, transfer, and capture of 'value' within the built environment," *Environment and Planning A* 24: 215–241.

—— (1992b) "Adelaide's heart transplant, 1970–88: 2 the 'transfer' of value within the housing market," *Environment and Planning A* 24: 323–339.

—— (1993) "Notwithstanding the exaggerated claims, residential revitalization really is changing the form of some Western cities: a response to Bourne," *Urban Studies* 30: 191–195.

—— (1995) "Building upon the foundations of gentrification: inner city housing development in Australia in the 1990s," *Urban Geography* 16: 70–90.

Bagli, C. V. (1991) "'De-gentrification' can hit when boom goes bust," *New York Observer* August 5–12.

Bailey, B. (1990) "The changing urban frontier: an examination of the meanings and conflicts of adaptation," MA thesis, Edinburgh University.

Baker, H. A., Jr. (1987) *Modernism and the Harlem Renaissance*, Chicago: University of Chicago Press.

Ball, M. A. (1979) "Critique of urban economics," *International Journal of Urban and Regional Research* 3: 309–332.

Baltimore City Department of Housing and Community Development (1977) *Homesteading: The Third Year, 1976*, Baltimore: Department of Housing and Community Development.

Baltzell, D. (1958) *Philadelphia Gentleman*, Glencoe: The Free Press.

Banfield, E. C. (1968) *The Unheavenly City: The Nature and Future of Our Urban Crisis*, Boston: Little, Brown.

Barrett, W. (1995) "Rudy's shrink rap," *Village Voice* May 9.

Barry, J. and Derevlany, J. (1987) *Yuppies Invade My House at Dinnertime*, Hoboken: Big River Publishing.

Bartelt, D. (1979) "Redlining in Philadelphia: an analysis of home mortgages in the Philadelphia area," mimeograph, Institute for the Study of Civic Values, Temple University.

Barthes, R. (1972) *Mythologies*, New York: Hill and Wang.

Baudelaire, C. (1947) *Paris Spleen*, New York: New Directions.

Beauregard, R. (1986) "The chaos and complexity of gentrification," in N. Smith and P. Williams (eds.) *Gentrification of the City*, Boston: Allen and Unwin.

—— (1989) *Economic Restructuring and Political Response*, Urban Affairs Annual Reviews 34, Newbury Park, Calif.: Sage Publications.

—— (1990) "Trajectories of neighborhood change: the case of gentrification," *Environment and Planning A* 22: 855–874.

—— (1993) *Voices of Decline: The Postwar Fate of US Cities*, Oxford: Basil Blackwell.

Bell, D. (1973) *The Coming of Post-industrial Society*, New York: Basic Books.

Beluszky, P. and Timár, J. (1992) "The changing political system and urban restructuring in Hungary," *Tijdschrift voor Economische en Sociale Geografie* 83: 380–389.

Bennet, J. (1992) "One emerald shines, others go unpolished," *New York Times* August 30.

Bennetts, L. (1982) "16 tenements to become artist units in city plan," *New York Times*, May 4.

Berman, M. (1982) *All That Is Solid Melts into Air: The Experience of Modernity*, New York: Simon and Schuster.

Bernstein, E. M. (1994) "A new Bradhurst," *New York Times* January 6.

Bernstein, R. (1990) "Why the cutting edge has lost its bite," *New York Times*, September 30.

Berry, B. (1973) *The Human Consequences of Urbanization*, London: Macmillan.

—— (1980) "Inner city futures: an American dilemma revisited," *Transactions of the Institute of British Geographers* NS 5, 1: 1–28.

—— (1985) "Islands of renewal in seas of decay," in P. Paterson (ed.) *The New Urban Reality*, Washington, DC: Brookings Institution.

Blumenthal, R. (1994) "Tangled ties and tales of FBI messenger," *New York Times*, January 9.

Boddy, M. (1980) *The Building Societies*, Basingstoke: Macmillan.

Bondi, L. (1991a) "Gender divisions and gentrification: a critique," *Transactions of the Institute of British Geographers* 16: 290–298.

—— (1991b) "Women, gender relations and the inner city," in M. Keith and A. Rogers (eds.) *Hollow Promises: Rhetoric and Reality in the Inner City*, London: Mansell.

Bontemps, A. (1972) *The Harlem Renaissance Remembered*, New York: Dodd, Mead and Co.

Bourassa, S. (1990) "On 'An Australian view of the rent gap' by Badcock," *Annals of the Association of American Geographers* 80: 458–459.

—— (1993) "The rent gap debunked," *Urban Studies* 30: 1731–1744.

Bourne, L. S. (1993) "The demise of gentrification? A commentary and prospective view," *Urban Geography* 14: 95–107.

Bowler, A. E. and McBurney, B. (1989) "Gentrification and the avante garde in New York's East Village: the good, the bad, and the ugly," paper presented at the annual conference of the American Sociological Association, San Francisco, August.

Boyle, M. (1992) "The cultural politics of Glasgow, European City of Culture: making sense of the role of the local state in urban regeneration," unpublished Ph.D. dissertation, Edinburgh University.

—— (1995) 'Still top our the agenda? Neil Smith and the reconciliation of capital and consumer approaches to the explanation of gentrification.' *Scottish Geographical Magazine* 111: 120–123.

Bradford, C. and Rubinowitz, L. (1975) "The urban–suburban investment–disinvestment process: consequences for older neighborhoods," *Annals of the American Academy of Political and Social Science* 422: 77–86.

Bridge, G. (1994) "Gentrification, class and residence: a reappraisal," *Environment and Planning D: Society and Space* 12: 31–51.

—— (1995) "The space for class? On class analysis in the study of gentrification," *Transactions of the Institute of British Geographers* NS 20, 2: 236–247.

Bronner, E. (1962) *William Penn's Holy Experiment*, Philadelphia: Temple University Publications.

Brown, J. (1973) "The whiting of Society Hill: black families refuse eviction," *Drummer* February 13.

Brown, P. L. (1990) "Lauren's wink at the wild side," *New York Times* February 8.

Bruce-Briggs, B. (1979) *The New Class?*, New Brunswick, N.J.: Transaction Books.

Bukharin, N. (1972 edn.) *Imperialism and World Economy*, London: Merlin.

Burt, N. (1963) *The Perennial Philadelphians*, London: Dent and Son.

Butler, S. (1981) *Enterprise Zones: Greenlining the Inner City*, New York: Universe Books.

Caris, P. (1996) "Declining suburbs: disinvestment in the inner suburbs of Camden County, New Jersey," unpublished Ph.D. dissertation, Rutgers University.

Carmody, D. (1984) "New day is celebrated for Union Square Park," *New York Times* April 20.

Carpenter, J. and Lees, L. (1995) "Gentrification in New York, London and Paris: an international comparison," *International Journal of Urban and Regional Research* 19: 286–303.

Carr, C. (1988) "Night clubbing: reports from the Tompkins Square Police Riot," *Village Voice* August 16.

Carroll, M. (1983) "A housing plan for artists loses in Board of Estimates," *New York Times* February 11.

Castells, M. (1976) "The Wild City," *Kapitalistate* 4–5: 2–30.

—— (1983) *The City and the Grassroots*, Berkeley: University of California Press.

—— (ed.) (1985) *High Technology, Space and Society*, volume 28, Urban Affairs Annual Reviews, London: Sage Publications.

Castillo, R. (1993) "A fragmentato da terra. Propriedade fundisria absoluta e espaço mercadoria no municipio de São Paulo," unpublished Ph.D. dissertation, Universidade de São Paulo.

Castrucci, A. *et al. YOUЯ HOUSE IS MINE* (1992) , New York: Bullet Space.

Caulfield, J. (1989) "'Gentrification' and desire," *Canadian Review of Sociology and Anthropology* 26, 4: 617–632.

—— (1994) *City Form and Everyday Life: Toronto's Gentrification and Critical Social Practice*, Toronto: University of Toronto Press.

Chall, D. (1984) "Neighborhood changes in New York City during the 1970's: are the gentry returning?," *Federal Reserve Bank of New York Quarterly Review* Winter: 38–48.

Charyn, J. (1985) *War Cries over Avenue C*, New York: Donald I. Fine, Inc.

Checkoway, B. (1980) "Large builders, federal housing programmes, and postwar suburbanization," *International Journal of International and Regional Research* 4: 21–44.

Chouinard, V., Fincher, R. and Webber, M. (1984) "Empirical research in scientific human geography," *Progress in Human Geography*, 8, 3: 347–380.

City of New York, Commission on Human Rights (1983) "Mortgage activity reports'.

City of New York, Department of City Planning (1981) "Sanborn vacant buildings file."

—— (1983) "Housing database: public and publicly aided housing."

City of New York, Harlem Task Force (1982) "Redevelopment strategy for Central Harlem."

Clark, E. (1987) *The Rent Gap and Urban Change: Case Studies in Malmö 1860–1985*, Lund: Lund University Press.

—— (1988) "The rent gap and transformation of the built environment: Case studies in Malmö 1860–1985," *Geografiska Annaler* 70B: 241–254.

—— (1991a) "Rent gaps and value gaps: complementary or contradictory," in J. van Weesep and S. Musterd (eds.) *Urban Housing for the Better-off: Gentrification in Europe*, Utrecht: Stedelijke Netwerken.

—— (1991b) "On gaps in gentrification theory," *Housing Studies* 7: 16–26.

—— (1992) "On blindness, centrepieces and complementarity in gentrification theory," *Transactions of the Institute of British Geographers* NS 17: 358–362.

—— (1994) "Toward a Copenhagen interpretation of gentrification," *Urban Studies* 31, 7: 1033–1042.

—— (1995) "The rent gap re-examined," *Urban Studies* 32: 1489–1503.

Clark, E. and Gullberg, A. (1991) "Long swings, rent gaps and structures of building provision: the postwar transformation of Stockholm's inner city," *International Journal of Urban and Regional Research* 15, 4: 492–504.

Claval, P. (1981) *La logique des villes*, Paris: Librairies Techniques.

Clay, P. (1979a) *Neighborhood Renewal*, Lexington, Mass.: D. C. Heath.

—— (1979b) *Neighborhood Reinvestment without Displacement: A Handbook for Local Government*, Cambridge, Mass.: Department of Urban Studies and Planning, Massachusetts Institute of Technology.

Connell, J. (1976) *The End of Tradition: Country Life in Central Surrey*, London: Routledge.

Coombs, O. (1982) "The new battle for Harlem," *New York* January 25.

Cortie, C. and van de Ven, J. (1981) "'Gentrification': keert de woonelite terug naar de stad?," *Geografisch Tijdschrift* 15: 429–446.

Cortie, C. and van Engelsdorp Gastelaars, R. (1985) "Amsterdam: decaying city, gentrifying city," in P. E. White and B. van der Knaap (eds.) *Contemporary Studies of Migration*, Norwich: Geo Books, International Symposia Series.

Cortie, C., van de Ven, J. and De Wijis-Mulkens (1982) "'Gentrification' in de Jordaan: de opkomst van een nieuwe birinenstadselite," *Geografisch Tijdschrift* 16: 352–379.

Cortie, C., Kruijt, B. and Musterd, S. (1989) 'Housing market change in Amsterdam: some trends," *Netherlands Journal of Housing and Environmental Research* 4: 217–233.

Counsell, G. (1992) "When it pays to be a vandal," *Independent on Sunday* June 21.

Crilley, D. (1993) "Megastructures and urban change: aesthetics, ideology and design," in P. Knox (ed.) *The Restless Urban Landscape*, Englewood Cliffs, NJ: Prentice-Hall.

Cybriwsky, R. (1978) "Social aspects of neighborhood change," *Annals of the Association of American Geographers* 68: 17–33.

—— (1980) "Historical evidence of gentrification," unpublished MS, Department of Geography, Temple University.

Dangschat, J. (1988) "Gentrification: der Wandel innenstadtnaher Nachbarschaften," *Kölner Zeitschrift für Soziologie und Sozialpsychologie, Sonderband*.

—— (1991) "Gentrification in Hamburg," in J. van Weesep and S. Musterd (eds.), *Urban Housing for the Better-Off: Gentrification in Europe*, Utrecht: Stedelijke Netwerken.

Daniels, L. (1982) "Outlook for revitalization of Harlem," *New York Times*, February 12.
—— (1983a) "Hope and suspicion mark plan to redevelop Harlem," *New York Times*, February 6.
—— (1983b) "Town houses in Harlem attracting buyers," *New York Times*, August 21.
—— (1984) "New condominiums at Harlem edge," *New York Times*, February 19.
Davis, J. T. (1965) "Middle class housing in the central city," *Economic Geography* 41: 238–251.
Davis, M. (1991) *City of Quartz*, London: Verso.
DeGiovanni, F. (1983) "Patterns of change in housing market activity in revitalizing neighborhoods," *Journal of the American Planning Association* 49: 22–39.
—— (1987) *Displacement Pressures in the Lower East Side*, working paper, Community Service Society of New York.
DePalma, A. (1988) "Can City's plan rebuild Lower East Side?," *New York Times* October 14.
Deutsche, R. (1986) "Krzysztof Wodiczko's *Homeless Projection* and the site of urban 'revitalization'," *October* 38: 63–98.
Deutsche, R. and Ryan, C. G. (1984) "The fine art of gentrification," *October* 31: 91–111.
Dieleman, F. M. and van Weesep, J. (1986) "Housing under fire: budget cuts, policy adjustments and market changes," *Tijdschrift voor Economische en Sociale Geografie* 77: 310–315.
"Disharmony and Housing" (1985) *New York Times* October 22.
Douglas, C. C. (1985) "149 win in auction of Harlem houses," *New York Times* August 17.
—— (1986) "Harlem warily greets plans for development," *New York Times* January 19.
Douglas, P. (1983) "Harlem on the auction block," *Progressive March*: 33–37.
Dowd, M. (1993) "The WASP descendancy," *New York Times Magazine* October 31, 46–48.
Downie, L. (1974) *Mortgage on America*, New York: Praeger.
Downs, A. (1982) "The necessity of neighborhood deterioration," *New York Affairs* 7, 2: 35–38.
Dunford, M. and Perrons, D. (1983) *The Arena of Capital*, London: Macmillan.
Edel, M. and Sclar, E. (1975) "The distribution of real estate value changes: metropolitan Boston, 1870–1970," *Journal of Urban Economics* 2: 366–387.
Egan, T. (1993) "In 3 progressive cities, stern homeless policies," *New York Times* December 12.
Ehrenreich, B. and Ehrenreich, J. (1979) "The professsional-managerial class," in P. Walker (ed.) *Between Labor and Capital*, Boston: South End Press.
Ekland-Olson, S., Kelly, W. R. and Eisenberg, M. (1992) "Crime and incarceration: some comparative findings from the 1980s," *Crime and Delinquency* 38: 392–416.
Engels, B. (1989) "The gentrification of Glebe: the residential restructuring of an inner Sydney suburb, 1960 to 1986," unpublished Ph.D. dissertation, University of Sydney.
Engels, F. (1975 edn.) *The Housing Question*, Moscow: Progress Publishers.
Fainstein, N. and Fainstein, S. (1982) "Restructuring the American city: a comparative perspective," in N. Fainstein and S. Fainstein (eds.) *Urban Policy under Capitalism*, Beverly Hills: Sage Publications.
Fainstein, S. (1994) *The City Builders: Property, Politics and Planning in London and New York*, Oxford: Basil Blackwell.
Fainstein, S., Harloe, M. and Gordon, I. (eds.) (1992) *Divided Cities: New York and London in the Contemporary World*, Oxford: Basil Blackwell.
Ferguson, S. (1988) "The boombox wars," *Village Voice* August 16.
—— (1991a) "Should Tompkins Square be like Gramercy?," *Village Voice* June 11.
—— (1991b) "The park is gone," *Village Voice* June 18.
Filion, P. (1991) "The gentrification–social structure dialectic: a Toronto case study," *International Journal of Urban and Regional Research* 15, 4: 553–574.

Firey, W. (1945) "Sentiment and symbolism as ecological variables," *American Sociological Review* 10: 140–148.

Fisher, I. (1993) "For homeless, a last haven is demolished," *New York Times* August 18.

Fitch, R. (1988) "What's left to write?: media mavericks lose their touch," *Voice Literary Supplement* May 19.

—— (1993) *The Assassination of New York*, New York: Verso.

Fodarero, L. (1987) "ABC's of conversion: 21 loft condos," *New York Times* March 22.

Foner, P. S. (1978) *The Labor Movement in the United States* volume 1, New York: International Publishers.

"$14.5 million arts project for Harlem" (1984) *Amsterdam News*, January 21.

Franco, J. (1985) "New York is a third world city," *Tabloid* 9: 12–19.

Friedrichs, J. (1993) "A theory of urban decline: economy, demography and political elites," *Urban Studies* 30, 6: 907–917.

Gaillard, J. (1977) *Paris: La Ville*. Paris: H. Champion.

Gale, D. E. (1976) "The back-to-the-city movement . . . or is it?," occasional paper, Department of Urban and Regional Planning, George Washington University.

—— (1977) "The back-to-the-city movement revisited," occasional paper, Department of Urban and Regional Planning, George Washington University.

Galster, G. C. (1987) *Homeowners and Neighborhood Reinvestment*, Durham, N.C.: Duke University Press.

Gans, H. (1968) *People and Plans*, New York: Basic Books.

Garreau, J. (1991) *Edge City: Life on the Frontier*, New York: Doubleday.

Gertler, M. (1988) "The limits to flexibility: comments on the post-Fordist vision of production and its geography," *Transactions of the Institute of British Geographers* 16: 419–422.

Gevirtz, L. (1988) "Slam dancer at NYPD," *Village Voice*, September 6.

Giddens, A. (1981) *A Contemporary Critique of Historical Materialism*. Volume 1: *Power, Property and the State*, Berkeley: University of California Press.

Gilmore, R. (1993) "Terror austerity race gender excess theater," in B. Gooding-Williams (ed.) *Reading Rodney King/Reading Urban Uprising*, New York: Routledge.

—— (1994) "Capital, state and the spatial fix: imprisoning the crisis at Pelican Bay," unpublished paper, Rutgers University.

Glass, R. (1964) *London: Aspects of Change*, London: Centre for Urban Studies and MacGibbon and Kee.

Glazer, L. (1988) "Heavenly developers building houses for the poor rich?," *Village Voice* October 11.

Goldberg, J. (1994) "The decline and fall of the Upper West Side: how the poverty industry is ripping apart a great New York neighborhood," *New York* April 25: 37–42.

Goldstein, R. (1983) "The gentry comes to the East Village," *Village Voice*, May 18.

Gooding-Williams, R. (ed.) (1993) *Reading Rodney King/Reading Urban Uprising*, New York: Routledge.

Goodwin, M. (1984) "Recovery making New York city of haves and have-nots," *New York Times* August 28.

Gottlieb, M. (1982) "Space invaders: land grab on the Lower East Side," *Village Voice* December 14.

Gould, A. (1981) "The salaried middle class in the corporatist welfare state," *Policy and Politics* 9: 4.

Gramsci, A. (1971 edn.) *Prison Notebooks*, New York: International Publishers.

Grant, L. (1990) "From riots to riches," *Observer Magazine* October 7.

Greenfield, A. M. and Co., Inc. (1964) "New town houses for Washington Square East: a technical report on neighborhood conservation," prepared for the Redevelopment Authority of Philadelphia.

Gutman, H. (1965) "The Tompkins Square 'riot' in New York City on January 13, 1874: a re-examination of its causes and consequences," *Labor History* 6: 44–70.

Hamnett, C. (1973) "Improvement grants as an indicator of gentrification in inner London," *Area* 5: 252–261.

—— (1984) "Gentrification and residential location theory: a review and assessment," in D. T. Herbert and R. J. Johnston (eds.) *Geography and the Urban Environment: Progress in Research and Applications*, vol. VI, New York: Wiley.

—— (1990) "London's turning," *Marxism Today* July, 26–31.

—— (1991) "The blind men and the elephant: the explanation of gentrification," *Transactions of the Institute of British Geographers* NS 16: 173–189.

—— (1992) "Gentrifiers or lemmings? A response to Neil Smith," *Transactions of the Institute of British Geographers* NS 17: 116–119.

Hamnett, C. and Randolph, W. (1984) "The role of landlord disinvestment in housing market transformation: an analysis of the flat break-up market in central London," *Transactions of the Institute of British Geographers* NS 9: 259–279.

—— (1986) "Tenurial transformation and the flat break-up market in London: the British condo experience," in N. Smith and P. Williams (eds.) *Gentrification of the City*, Boston: Allen and Unwin.

Hampson, R. (1982) "Will whites buy the future of Harlem?," *Record* July.

Haraway, D. and Harvey, D. (1995) "Nature, politics and possibilities: a discussion and debate with David Harvey and Donna Haraway," *Environment and Planning D: Society and Space* 13: 507–528.

Harlem Urban Development Corporation (1982) "Analyses of property sales within selected areas of the Harlem UDC task force area," revised edition.

Harloe, M. (1984) "Sector and class: a critical comment," *International Journal of Urban and Regional Research* 8, 2: 228–237.

Harman, C. (1981) "Marx's theory of crisis and its critics," *International Socialism* 11: 30–71.

Harris, N. (1980a) "Crisis and the core of the world system," *International Socialism* 10: 24–50.

—— (1980b) "Deindustrialization," *International Socialism* 7: 72–81.

—— (1983) *Of Bread and Guns: The World Economy in Crisis*, New York: Penguin.

Hartman, C. (1979) "Comment on 'Neighborhood revitalization and displacement: a review of the evidence'," *Journal of the American Planning Association* 45, 4: 488–491.

Harvey, D. (1973) *Social Justice and the City*, Baltimore: Johns Hopkins University Press.

—— (1974) "Class monopoly rent, finance capital and the urban revolution," *Regional Studies* 8: 239–255.

—— (1975) "Class structure in a capitalist society and the theory of residential differentiation," in M. Chisholm and R. Peel (eds.) *Processes in Physical and Human Geography*, Edinburgh: Heinemann.

—— (1977) "Labor, capital and class struggle around the built environment in advanced capitalist societies," *Politics and Society* 7: 265–275.

—— (1978) "The urban process under capitalism: a framework for analysis," *International Journal of Urban and Regional Research* 2, 1: 100–131.

—— (1982) *The Limits to Capital*, Oxford: Basil Blackwell.

—— (1985a) *Consciousness and the Urban Experience: Studies in the History and Theory of Capitalist Urbanization*, Oxford: Basil Blackwell.

—— (1985b) *The Urbanization of Capital: Studies in the History and Theory of Capitalist Urbanization*, Oxford: Basil Blackwell.

—— (1989) *The Condition of Post-modernity*, Oxford: Blackwell.

Harvey, D., Chaterjee, L., Wolman, M. and Newman, J. (1972) *The Housing Market and Code Enforcement in Baltimore*, Baltimore: City Planning Department.

Harvey, D. and Chaterjee, L. (1974) "Absolute rent and the structuring of space by governmental and financial institutions," *Antipode* 6, 1: 22–36.

Hegedüs, J. and Tosics, I. (1991) "Gentrification in Eastern Europe: the case of Budapest," in J. van Weesep and S. Musterd (eds.) *Urban Housing for the Better-Off: Gentrification in Europe*, Utrecht: Stedelijke Netwerken.

—— (1993) "Changing public housing policy in central European metropolis: the case of Budapest." Paper presented to European Network for Housing Research Conference, Budapest, September 7–10.

Heilbrun, J. (1974) *Urban Economics and Public Policy*, New York: St. Martin's Press.

Henwood, D. (1988) "Subsidizing the rich," *Village Voice* August 30.

Hoch, C. and Slayton, R. A. (1989) *New Homeless and Old: Community and the Skid Row Hotel*, Philadelphia: Temple University Press.

"Home sales low in '82, but a recovery is seen" (1983) *New York Times* February 1.

Hoyt, H. (1933) *One Hundred Years of Land Values in Chicago*, Chicago: University of Chicago Press.

Huggins, N. R. (1971) *Harlem Renaissance*, London: Oxford University Press.

Ingersoll, A. C. (1963) "A Society Hill restoration," *Bryn Mawr Alumnae Bulletin* Winter.

International Labour Organisation (1994) *Year Book of Labour Statistics*, Geneva: ILO.

Jack, I. (1984) "The repackaging of Glasgow," *Sunday Times Magazine* December 2.

Jackson, P. (1985) "Neighbourhood change in New York: the loft conversion process," *Tijdschrift voor Economische en Sociale Geografie* 76, 3: 202–215.

Jacobs, J. (1961) *The Life and Death of Great American Cities*, New York: Random House.

Jager, M. (1986) "Class definition and the aesthetics of gentrification: Victoriana in Melbourne," in N. Smith and P. Williams (eds.) *Gentrification of the City*, Boston: Allen and Unwin.

James, F. (1977) "Private reinvestment in older housing and older neighborhoods: recent trends and forces," Committee on Banking, Housing and Urban Affairs, US Senate, July 7 and 8, Washington, DC.

Jencks, C. (1994a) "The homeless," *New York Review of Books* April 21: 20–27.

—— (1994b) "Housing the homeless," *New York Review of Books* May 12: 39–46.

—— (1994c) *The Homeless*, Cambridge, Mass.: Harvard University Press.

Jobse, R. B. (1987) "The restructuring of Dutch cities," *Tijdschrift voor Economische en Sociale Geografie* 78: 305–311.

Kaplan, E. (1994) "Streetwatch, redux," *Village Voice* June 21.

Kary, K. (1988) "The gentrification of Toronto and the rent gap theory," in T. Bunting and P. Filion (eds.) *The Changing Canadian Inner City*, Department of Geography Publication 31, University of Waterloo.

Katz, C. (1991a) "An agricultural project comes to town: consequences of an encounter in the Sudan," *Social Text* 28, 31–38.

—— (1991b) "Sow what you know: the struggle for social reproduction in rural Sudan," *Annals of the Association of American Geographers* 81, 488–514.

—— (1991c) "A cable to cross a curse," unpublished paper.

Katz, C. and Smith, N. (1992) "LA intifada: interview with Mike Davis," *Social Text* 33: 19–33.

Katz, S. and Mayer, M. (1985) "Gimme shelter: self-help housing struggles within and against the state in New York City and West Berlin," *International Journal of Urban and Regional Research* 9: 15–17.

Katznelson, I. (1981) *City Trenches: Urban Politics and the Patterning of Class in the United States*, Chicago: University of Chicago Press.

Kay, H. (1966) "The industrial corporation in urban renewal," in J. Q. Wilson (ed.) *Urban Renewal*, Cambridge, Mass.: MIT Press.

Kendig, H. (1979) "Gentrification in Australia," in J. J. Palen and B. London (eds.) *Gentrification, Displacement and Neighborhood Revitalization*, Albany: State University of New York Press.

Kifner, J. (1991) "New York closes park to homeless," *New York Times* June 4.

Knopp, L. (1989) "Gentrification and gay community development in a New Orleans neighborhood," Ph.D. dissertation, Department of Geography, University of Iowa.

—— (1990a) "Some theoretical implications of gay involvement in an urban land market," *Political Geography Quarterly* 9: 337–352.

—— (1990b) "Exploiting the rent gap: the theoretical significance of using illegal appraisal schemes to encourage gentrification in New Orleans," *Urban Geography* 11: 48–64.

Koptiuch, K. (1991) "Third-worlding at home," *Social Text* 28: 87–99.

Kovács, Z. (1993) "Social and economic transformation in Budapest," paper presented to European Network for Housing Research Conference, Budapest, September 7–10.

—— (1994) "A city at the crossroads: social and economic transformation in Budapest," *Urban Studies* 31: 1081–1096.

Kraaivanger, H. (1981) "The Battle of Waterlooplein," *Move* 3: 4.

Kruger, K.-H. (1985) "Oh, baby. Scheisse. Wie ist das gekommen?," *Der Spiegel* March 11.

Lake, R. W. (1979) *Real Estate Tax Delinquency: Private Disinvestment and Public Response*, Piscataway, N.J.: Center for Urban Policy Research, Rutgers University.

Lamarche, F. (1976) "Property development and the economic foundations of the urban question," in C. G. Pickvance (ed.) *Urban Sociology: Critical Essays*, London: Methuen.

Laska, S. and Spain, D. (eds.) (1980) *Back to the City: Issues in Neighborhood Renovation*, Elmsford, N.Y.: Pergamon Press.

Laurenti, L. (1960) *Property Values and Race*, Berkeley: University of California Press.

Lauria, M. and Knopp, L. (1985) "Toward an analysis of the role of gay communities in the urban renaissance," *Urban Geography* 6: 152–169.

Lee, E. D. (1981) "Will we lose Harlem? The symbolic capital of Black America is threatened by gentrification," *Black Enterprise* June: 191–200.

Lees, L. (1989) "The gentrification frontier: a study of the Lower East Side area of Manhattan, New York City," unpublished BA dissertation, Queen's University, Belfast.

—— (1994) "Gentrification in London and New York: an Atlantic gap?" *Housing Studies* 9, 2: 199–217.

—— (1994) "Rethinking gentrification: beyond the positions of economics or culture," *Progress in Human Geography* 18, 2: 137–150.

Lees, L. and Bondi, L. (1995) "De/gentrification and economic recession: the case of New York City," *Urban Geography* 16: 234–253.

"Les Africains sont les plus mal logés" (1992) *Libération* September 12.

"Les sans-abri du XIIIe vont être délogés" (1991) *Journal du Dimanche* September 1.

"Les sans-logis de Vincennes laissent le camp aux mal-logés" (1992) *Libération* September 5–6.

Levin, K. (1983) "The neo-frontier," *Village Voice* January 4.

Levy, P. (1978) "Inner city resurgence and its societal context," paper presented at the annual conference of Association of American Geographers, New Orleans.

Lewis, D. (1981) *When Harlem Was in Vogue*, New York: Random House.

Ley, D. (1978) "Inner city resurgence and its social context," paper presented at the annual conference of the Association of American Geographers, New Orleans.

—— (1980) "Liberal ideology and the postindustrial city," *Annals of the Association of American Geographers* 70: 238–258.

—— (1986) "Alternative explanations for inner-city gentrification," *Annals of the Association of American Geographers* 76: 521–535.

—— (1992) "Gentrification in recession: social change in six Canadian inner cities, 1981–1986," *Urban Geography* 13, 3: 230–256.

Limerick, P. N. (1987) *The Legacy of Conquest: The Unbroken Past of the American West*, New York: Norton.

Linton, M. (1990) "If I can get it for £148 why pay more?" *Guardian* 16 May 1990.

Lipton, S. G. (1977) "Evidence of central city revival," *Journal of the American Institute of Planners* 43: 136–147.

Long, L. (1971) "The city as reservation," *Public Interest* 25: 22–38.

"Los Angeles plans a camp for downtown homeless" (1994) *International Herald Tribune*, October 15–16.

Lowenthal, D. (1986) *The Past Is a Foreign Country*, Cambridge: Cambridge University Press.

Lowry, I. S. (1960) "Filtering and housing costs: a conceptual analysis," *Land Economics* 36: 362–370.

"Ludlow Street" (1988) *New Yorker* February 8.

MacDonald, G. M. (1993) "Philadelphia's Penn's Landing: changing concepts of the central river front," *Pennsylvania Geographer* 31, 2: 36–51.

McDonald, J. F. and Bowman, H. W. (1979) "Land value functions: a reevaluation," *Journal of Urban Economics* 6: 25–41.

McGhie, C. (1994) "Up and coming but never arrived," *Independent on Sunday* March 27.

McKay, C. (1928) *Home to Harlem*, New York: Harper and Row.

"Make Tompkins Square a park again" (1991) *New York Times* May 31.

Mandel, E. (1976) "Capitalism and regional disparities," *South West Economy and Society* 1: 41–47.

Marcuse, P. (1984) "Gentrification, residential displacement and abandonment in New York City," report to the Community Services Society.

—— (1986) "Abandonment, gentrification and displacement: the linkages in New York City," in N. Smith and P. Williams (eds.) *Gentrification of the City*, Boston: Allen and Unwin.

—— (1988) "Neutralizing homelessness," *Socialist Review* 88, 1: 69–96.

—— (1991) "In defense of gentrification," *Newsday* December 2.

Markusen, A. (1981) "City spatial structure, women's household work, and national urban policy," in C. Stimpson, E. Dixler, M. J. Nelson and K. B. Yatrakis (eds.) *Women and the City*, Chicago: University of Chicago Press.

Martin, D. (1993) "Harlem landlord sees dream of affordable housing vanish," *New York Times* November 7.

Marx, K. (1963 edn.) *The 18th Brumaire of Louis Bonaparte*, New York: International Publishers.

—— (1967 edn.) *Capital* (three volumes), New York: International Publishers.

—— (1973 edn.) *Grundrisse*, London: Pelican.

—— (1974 edn.) *The Civil War in France*, Moscow: Progress Publishers.

Massey, D. (1978) "Capital and locational change: the UK electrical engineering and electronics industry," *Review of Radical Political Economics* 10, 3: 39–54.

Massey, D. and Meegan, R. (1978) "Industrial restructuring versus the cities," *Urban Studies* 15: 273–288.

McGee, H. W., Jr. (1991) "Afro-American resistance to gentrification and the demise of integrationist ideology in the United States," *Urban Lawyer* 23: 25–44.

Miller, S. (1965) "The 'New' Middle Class," in A. Shostak and W. Gomberg (eds.) *Blue-Collar Worker*, Englewood Cliffs, N.J.: Harper and Row.

Mills, C. A. (1988) "'Life on the upslope': the postmodern landscape of gentrification," *Environment and Planning D: Society and Space* 6: 169–189.

Mills, E. (1972) *Studies in the Structure of Urban Economy*, Baltimore: Johns Hopkins University Press.

Mingione, E. (1981) *Social Conflict and the City*, Oxford: Basil Blackwell.

Mitchell, D. (1995a) "The end of public space? People's Park, definitions of the public, and democracy," *Annals of the Association of American Geographers* 85, 1: 108–133.

—— (1995b) "There's no such thing as culture: towards a reconceptualization of the idea of culture in geography," *Transactions of the Institute of British Geographers* NS 20: 102–116.

Mollenkopf, J. and Castells, M. (eds.) (1991) *Dual City*, New York: Russell Sage Foundation.

Morgan, T. (1991) "New York City bulldozes squatters' shantytowns," *New York Times* October 16.

Morris, A. E. J. (1975) "Philadelphia: idea powered planning," *Built Environment Quarterly* 1: 148–152.

Moufarrege, N. (1982) "Another wave, still more savagely than the first: Lower East Side, 1982," *Arts* 57, 1: 73.

—— (1984) "The years after," *Flash Art* 118: 51–55.

Mullins, P. (1982) "The 'middle-class' and the inner city," *Journal of Australian Political Economy* 11: 44–58.

Musterd, S. (1989) "Upgrading and downgrading in Amsterdam neighborhoods," Instituut voor Sociale Geografie, Universitet van Amsterdam.

Musterd, S. and van de Ven, J. (1991) "Gentrification and residential revitalization," in J. van Weesep and S. Musterd (eds.) *Urban Housing for the Better-Off: Gentrification in Europe*, Utrecht: Stedelijke Netwerken.

Musterd, S. and J. van Weesep (1991) "European gentrification or gentrification in Europe?," in J. van Weesep and S. Musterd (eds.) *Urban Housing for the Better-Off: Gentrification in Europe*, Utrecht: Stedelijke Netwerken.

Murray, M. J. (1994) *The Revolution Deferred: The Painful Birth of Post-apartheid South Africa*, London: Verso.

Muth, R. (1969) *Cities and Housing*, Chicago: University of Chicago Press.

New York City Partnership (1987) "New homes program": announcement of the application period for the Towers on the Park.

Nicolaus, M. (1969) "Remarks at ASA convention," *American Sociologist* 4: 155.

Nitten (1992) *Christiania Tourist Guide*, Nitten: Copenhagen.

"Notes and comment" (1984) *New Yorker* September 24.

Nundy, J. (1991) "Homeless pawns in a party political game," *Independent on Sunday* August 11.

O'Connor, J. (1984) *Accumulation Crisis*, Oxford: Basil Blackwell.

Old Philadelphia Development Corporation (1970) "Annual report'.

Old Philadelphia Development Corporation (1975) "Statistics on Society Hill," unpublished report.

Oreo Construction Services (1982) "An analysis of investment opportunities in the East Village'.

Oser, A. S. (1985) "Mixed-income high-rise takes condominium form," *New York Times* June 30.

—— (1994) "Harlem rehabilitation struggle leaves casualties," *New York Times* December 4.

Osofsky, G. (1971) *Harlem: The Making of a Ghetto*, second edition New York: Harper and Row.

Owens, C. (1984) "Commentary: the problem with puerilism," *Art in America* 72, 6: 162–163.

Park, R. E., Burgess, E. W. and McKenzie, R. (1925) *The City*, Chicago: University of Chicago Press.

Parker, R. (1972) *The Myth of the Middle Class*, New York: Harper and Row.

Pei, I. M. and Associates (undated) "Society Hill, Philadelphia: a plan for redevelopment," prepared for Webb and Knapp.

Perlez, J. (1993) "Gentrifiers march on, to the Danube Banks," *New York Times* August 18.

Pickvance, C. (1994) "Housing privatisation and housing protest in the transition from state socialism: a comparative study of Budapest and Moscow," *International Journal of Urban and Regional Research* 18: 433–450.

Pinkney, D. H. (1957) *Napoleon III and the Rebuilding of Paris*, Princeton, N.J.: Princeton University Press.

Pitt, D. E. (1989) "PBA leader assails report on Tompkins Square melee," *New York Times* April 21.

Pitt, J. (1977) *Gentrification in Islington*, London: Peoples Forum.

Poulantzas, N. (1975) *Classes in Contemporary Capitalism*, London: New Left Books.

"Production group wants to air 'execution of month' on pay-TV" (1994) *San Francisco Chronicle*, April 4.

"Profile of a winning sealed bidder" (1985) *Harlem Entrepreneur Portfolio* Summer.

"Profiles in brownstone living" (1985) *Harlem Entrepreneur Portfolio* Summer.

Purdy, M. W. and Kennedy, S. G. (1995) "Behind collapse of a building, an 80's investment that did, too," *New York Times* March 26.

Queiroz Ribeiro, L. C. and Correa do Lago, L. (1995) "Restructuring in large Brazilian cities: the centre/periphery model," *International Journal of Urban and Regional Research* 19: 369–382.

Ranard, A. (1991) "An artists' oasis in Tokyo gives way to gentrification," *International Herald Tribune* January 4.

Ravo, N. (1992a) "New housing at lowest since '85," *New York Times* August 30.

—— (1992b) "Surge in home foreclosures and evictions shattering families," *New York Times* November 15.

Real Estate Board of New York, Inc., Research Department (1985) "Manhattan real estate open market sales, 1980–1984," mimeo.

Reaven, M. and Houk, J. (1994) "A hisotyr of Tomkins Square Park," in J. Abu-Lughod (ed.) *From Urban Village to East Village*, Cambridge, Mass.: Blackwell.

Redevelopment Authority of Philadelphia (undated) "Final project reports for Washington Square East urban renewal area, units 1, 2, and 3," unpublished, Philadelphia: Redevelopment Authority.

Regional Plan Association of America (1929) *New York Regional Plan*, New York: Regional Plan Association.

Reid, L. (1995) "Flexibilisation: past, present and future," *Scottish Geographical Magazine* 111, 1: 58–62.

Reiss, M. (1988) "Luxury housing opposed by community" *New Common Good*, July.

Rex, J. and Moore, R. (1967) *Race, Community and Conflict*, London: Oxford University Press.

Rickelfs, R. (1988) "The Bowery today: a skid row area invaded by yuppies," *Wall Street Journal* November 13.

Riesman, D. (1961) *The Lonely Crowd*, New Haven, Conn.: Yale University Press.

Riis, J. (1971) *How the Other Half Lives*, New York: Dover Publications.

Roberts, F. (1979) "Tales of the pioneers," *Philadelphia Inquirer* August 19.

Roberts, S. (1991a) "Crackdown on homeless and what led to shift," *New York Times* October 28.

—— (1991b) "Evicting the homeless," *New York Times* June 22.

Robinson, W. and McCormick, C. (1984) "Slouching toward Avenue D," *Art in America* 72, 6: 135–161.

Rodger, R. (1982) "Rents and ground rents: housing and the land market in nineteenth century Britain," in J. H. Johnson and C. G. Pooley (eds.) *The Structure of Nineteenth-Century Cities*, London: Croom Helm.

Rose, D. (1984) "Rethinking gentrification: beyond the uneven development of Marxist urban theory," *Environment and Planning D: Society and Space* 2: 47–74.

—— (1987) "Un aperçu féministe sur la réstructuration de l'emploi et sur la gentrification: le cas de Montréal," *Cahiers de Géographie du Québec* 31: 205–224.

Rose, David (1989) "The Newman strategy applied in 'frontline' tactics," *Guardian* August 31 1989.

Rose, H. M. (1982) "The future of black ghettos," in G. Gappert and R. Knight (eds.) *Cities in the 21st Century*, Urban Affairs Annual Reviews 23. Beverly Hills: Sage Publications.

Rose, J. and Texier, C. (eds.) (1988) *Between C & D: New Writing from the Lower East Side Fiction Magazine*, New York: Penguin.

Ross, A. (1994) "Bombing the Big Apple," in *The Chicago Gangster Theory of Life: Nature's Debt to Society*, London: Verso.

Rossi, P. (1989) *Down and Out in America: The Origins of Homelessness*, Chicago: University of Chicago Press.

Rothenberg, T. Y. (1995) "'And she told two friends . . .': lesbians creating urban social space," in D. Bell and G. Valentine (eds.) *Mapping Desire*, London: Routledge.

Routh, G. (1980) *Occupation and Pay in Great Britain 1906–1979*, London: Macmillan.

Roweis, S. and Scott, A. (1981) "The urban land question," in M. Dear and A. Scott (eds.) *Urbanization and Urban Planning in Capitalist Society*, New York: Methuen.

Rudbeck, C. (1994) "Apocalypse soon: how Miami nice turned to Miami vice in the eyes of crime writer Carl Hiaasen, who uncovers the dark side of the Sunshine State," *Scanorama* May: 52–59.

Rutherford, M. (1981) "Why Labour is losing more than a deposit," *Financial Times* November 28.

Rutkoff, P. M. (1981) *Revanche and Revision: The Ligue des Patroites and the Origins of the Radical Right in France, 1882–1900*. Athens: Ohio University.

Salins, P. (1981) "The creeping tide of disinvestment," *New York Affairs* 6, 4: 5–19.

Samuel, R. (1982) "The SDP and the new political class," *New Society* April 22: 124–127.

Sanders, H. (1980) "Urban renewal and the revitalized city: a reconsideration of recent history," in D. Rosenthal (ed.) *Urban Revitalization*, Beverly Hills: Sage Publications.

Sassen, S. (1988) *The Mobility of Labour and Capital*, Cambridge: Cambridge University Press.

——— (1991) *The Global City*, Princeton, N.J.: Princeton University Press.

Saunders, P. (1978) "Domestic property and social class," *International Journal of Urban and Regional Research* 2: 233–251.

——— (1981) *Social Theory and the Urban Question*, London: Hutchinson.

——— (1984) "Beyond housing classes: the sociological significance of private property rights and means of consumption," *International Journal of Urban and Regional Research* 8, 2: 202–227.

——— (1990) *A Nation of Homeowners*, London: Allen and Unwin.

Sayer, A. (1982) "Explanation in economic geography," *Progress in Human Geography* 6: 68–88.

Schemo, D. J. (1994) "Facing big-city problems, L.I. suburbs try to adapt," *New York Times* March 16.

Scott, A. (1981) "The spatial structure of metropolitan labor markets and the theory of intra-urban plant location," *Urban Geography* 2, 1: 1–30.

Scott, H. (1984) *Working Your Way to the Bottom: The Feminization of Poverty*, London: Pandora.

Séguin, A.-M. (1989) "Madame Ford et l'espace: lecture féministe de la suburbanisation," *Recherches Féministes* 2, 1: 51–68.

Servin, J. (1993) "Mall evolution," *New York Times* October 10.

Shaman, D. (1988) "Lower East Side buildings rehabilitated," *New York Times*, April 1.

Slotkin, R. (1985) *Fatal Environment: The Myth of the Frontier in the Age of Industrialization 1800–1890*, New York: Atheneum.

Smith, A. (1989) "Gentrification and the spatial contribution of the state: the restructuring of London's Docklands," *Antipode* 21, 3: 232–260.

Smith, M. P. (ed.) (1984) *Cities in Transformation: Class, Capital and the State*, Urban Affairs Annual Reviews 26, Newbury Park, Calif.: Sage Publications.

Smith, M. P. and Feagin, J. (eds.) (1996) *The Bubbling Cauldron*, Minneapolis: University of Minnesota Press.

Smith, N. (1979a) "Toward a theory of gentrification: a back to the city movement by capital not people," *Journal of the American Planning Association* 45: 538–548.

——— (1979b) "Gentrification and capital: theory, practice and ideology in Society Hill," *Antipode* 11, 3: 24–35.

——— (1982) "Gentrification and uneven development," *Economic Geography* 58: 139–155.

——— (1984) *Uneven Development: Nature, Capital and the Production of Space*, Oxford: Basil Blackwell.

——— (1986) "Gentrification, the frontier, and the restructuring of urban space," in N. Smith and P. Williams (eds.) *Gentrification of the City*, Boston: Allen and Unwin.

——— (1987) "Gentrification and the rent gap," *Annals of the Association of American Geographers* 77: 462–465.

—— (1991) "On gaps in our knowledge of gentrification," in J. van Weesep and S. Musterd (eds.) *Urban Housing for the Better-Off: Gentrification in Europe*, Utrecht: Stedelijke Netwerken.

—— (1992) "Blind man's bluff or, Hamnett's philosophical individualism in search of gentrification," *Transactions of the Institute of British Geographers* NS 17: 110–115.

—— (1995a) "Remaking scale: competition and cooperation in pre/postnational Europe," in H. Eskelinen and F. Snickars (eds.) *Competitive European Peripheries*, Berlin: Springer-Verlag.

—— (1995b) "Gentrifying Theory," *Scottish Geographical Magazine* 111: 124–126.

—— (1996a) "After Tompkins Square Park: degentrification and the Revanchist city," in A. King (ed.) *Re-presenting the City: Ethnicity, Capital and Culture in the 21st Century Metropolis*, London: Macmillan.

—— (1996b) "The production of nature," in G. Robertson and M. Mash (eds.) *FutureNatural*. London: Routledge.

—— (1996c) "The revanchist city: New York's homeless wars," *Polygraph* (forthcoming).

Smith, N. and LeFaivre, M. (1984), "A class analysis of gentrification," in B. London and J. Palen (eds.) *Gentrification, Displacement and Neighborhood Revitalization*. Albany: State University of New York Press.

Smith, N. and Williams, P. (eds.) (1986) *The Gentrification of the City*, Boston: Allen and Unwin.

Sohn-Rethel, A. (1978) *Intellectual and Manual Labor*, London: Macmillan.

Soja, E. (1980) "The socio-spatial dialectic," *Annals of the Association of American Geographers* 70: 207–225.

Squires, G. D., Velez, W. and Taeuber, K. E. (1991) "Insurance redlining, agency location and the process of urban disinvestment," *Urban Affairs Quarterly* 26, 4: 567–588.

Stallard, K., Ehrenreich, B. and Sklar, H. (1983) *Poverty in the American Dream: Women and Children First*, Boston: South End Press.

Stecklow, S. (1978) "Society Hill: rags to riches," *Evening Bulletin* January 13.

Stegman, M. A. (1972) *Housing Investment in the Inner City: The Dynamics of Decline*, Cambridge, Mass.: MIT Press.

—— (1982) *The Dynamics of Rental Housing in New York City*, New Brunswick, NJ: Center for Urban Policy Research, Rutgers University.

Steinberg, J., van Zyl, P. and Bond, P. (1992) "Contradictions in the transition from urban apartheid: barriers to gentrification in Johannesburg," in D. M. Smith (ed.) *The Apartheid City and Beyond: Urbanization and Social Change in South Africa*, London: Routledge.

Sternlieb, G. (1971) "The city as sandbox," *Public Interest* 25: 14–21.

Sternlieb, G. and Burchell, R. W. (1973) *Residential Abandonment: The Tenement Landlord Revisited*, Piscataway, N.J.: Center for Urban Policy Research, Rutgers University.

Sternlieb, G. and Hughes, J. (1983) "The uncertain future of the central city," *Urban Affairs Quarterly* 18, 4: 455–472.

Sternlieb, G. and Lake, R. W. (1976) "The dynamics of real estate tax delinquency," *National Tax Journal* 29: 261–271.

Stevens, W. K. (1991) "Early farmers and sowing of languages," *New York Times* May 9.

Stevenson, G. (1980) "The abandonment of Roosevelt Gardens," in R. Jensen (ed.) *Devastation/Reconstruction: The South Bronx*, New York: Bronx Museum of the Arts.

Stratton, J. (1977) *Pioneering in the Urban Wilderness*, New York: Urizen Books.

Sumka, H. (1979) "Neighborhood revitalization and displacement: a review of the evidence," *Journal of the American Planning Association* 45: 480–487.

Susser, I. (1993) "Creating family forms: the exclusion of men and teenage boys from families in the New York City shelter system, 1987–1992," *Critique of Anthropology* 13: 266–283.

Swart, P. (1987) "Gentrification as an urban phenomenon in Stellenbosch, South Africa," *Geo-Stell* 11: 13–18.

Swierenga, R. P. (1968) *Pioneers and Profits: Land Speculation on the Iowa Frontier*, Ames, IA: Iowa State University Press.

Sýkora, L. (1993) "City in transition: the role of rent gaps in Prague's revitalization," *Tijdschrift voor Economische en Sociale Geografie* 84: 281–293.

Taylor, M. M. (1991) "Home to Harlem: Black identity and the gentrification of Harlem," unpublished Ph.D. dissertation, Harvard University.

Toth, J. (1993) *The Mole People: Life in the Tunnels beneath New York City*, Chicago: Chicago Review Press.

Turner, F. J. (1958) *The Frontier in American History*, New York: Holt, Rinehart and Winston.

Unger, C. (1984) "The Lower East Side: there goes the neighborhood," *New York* May 28, 32–41.

United States Department of Commerce, Bureau of the Census (1972) *Census of Population and Housing. Census tracts, New York, NY SMSA, 1970*, Washington, DC.

—— (1983) *Census of Population and Housing. Census Tracts, New York, NY–NJ SMSA, 1980*, Washington, DC.

—— (1993) *Census of Population and Housing. Census Tracts, New York, NY–NJ PMSA, 1990*, Washington, DC.

—— (1994) *Statistical Abstract of the United States, 1994*, 114th Edition, Washington, DC.

United States Department of Commerce, Economic and Statistics Administration, Bureau of the Census (1993) *Money Income of Households, Families, and Persons in the United States: 1992*. Series P60–184, Washington, DC.

Urban Land Institute (1976) *New Opportunities for Residential Development in Central Cities*, Report no. 25, Washington, DC: Urban Land Institute.

Uzelac, E. (1991) "'Out of choices': urban pioneers abandon inner cities," *Sun* September 18.

Vance, T. N. (1951) "The permanent war economy," *New International* January–February.

van Kempen, R. and van Weesep, J. (1993) "Housing policy, gentrification and the urban poor: the case of Utrecht, the Netherlands," paper presented at the ENHR Conference on Housing Policy in Europe in the 1990s: Transformation in the East, Transference in the West, Budapest, September 7–10.

van Weesep, J. (1984) "Condominium conversion in Amsterdam: boon or burden," *Urban Geography* 5: 165–177.

—— (1986) *Condominium: A New Housing Sector in the Netherlands*, The Hague: CIP-Gegevens Koninklijke Bibliotheek.

—— (1988) "Regional and urban development in the Netherlands: the retreat of government," *Geography* 73: 97–104.

—— (1994) "Gentrification as a research frontier," *Progress in Human Geography* 18: 74–83.

van Weesep, J. and Maas, M. W. A. (1984) "Housing policy and conversions to condominiums in the Netherlands," *Environment and Planning A* 16: 1149–1161.

van Weesep, J. and Wiegersma, M. (1991) "Gentrification in the Netherlands: behind the scenes," in J. van Weesep and S. Musterd (eds.) *Urban Housing for the Better-Off: Gentrification in Europe*, Utrecht: Stedelijke Netwerken.

Vázquez, C. (1992) "Urban policies and gentrification trends in Madrid's inner city," *Netherlands Journal of Housing and Environmental Research* 7, 4: 357–376.

Vervaeke, M. and Lefebvre, B. (1986) *Habiter en Quartier Ancien*, Paris and Lille: CNRS.

Virilio, P. (1994) "Letter from Paris," *ANY Magazine* 4: 62.

Wagner, D. (1993) *Checkerboard Squares: Culture and Resistance in a Homeless Community*, Boulder, Colo.: Westview Press.

Walker, R. (1977) "The suburban solution," unpublished Ph.D. dissertation, Johns Hopkins University.

—— (1978) "The transformation of urban structure in the nineteenth century and the beginnings of suburbanization," in K. Cox (ed.) *Urbanization and Conflict in Market Societies*, Chicago: Maaroufa Press.

—— (1981) "A theory of suburbanization: capitalism and the construction of urban space in the United States," in M. Dear and A. J. Scott (eds.) *Urbanization and Urban Planning in Capitalist Society*, London: Methuen.

Walker, R. and Greenberg, D. (1982) "Post industrialism and political reform: a critique," *Antipode* 14, 1: 17–32.

Warde, A. (1991) "Gentrification as consumption: issues of class and gender," *Environment and Planning D: Society and Space* 9: 223–232.

Warner, S. B. (1972) *The Urban Wilderness: A History of the American City*, New York: Harper and Row.

Watson, S. (1986) "Housing and the family: the marginalization of non-family households in Britain," *International Journal of Urban and Regional Research* 10: 8–28.

Webber, M. (1963) "Order in diversity: community without propinquity," in L. Wingo (ed.) *Cities and Space: The Future Use of Urban Land*, Baltimore: Johns Hopkins University Press.

—— (1964a) "Culture, territoriality and the elastic mile," *Papers of the Regional Science Association* 13: 59–69.

—— (1964b) *The Urban Place and the Non-place Urban Realm: Explorations into Urban Structure*, Philadelphia: University of Pennsylvania Press.

Weinberg, B. (1990) "Is gentrification genocide? Squatters build an alternative vision for the Lower East Side," *Downtown 181*, February 14.

White, M. and White, L. (1977) *The Intellectual versus the City*, New York: Oxford University Press.

White, R. (1991) *Rude Awakenings: What the Homeless Crisis Tells Us*, San Francisco: Institute for Contemporary Studies Press.

Whitehand, J. (1972) "Building cycles and the spatial form of urban growth," *Transactions of the Institute of British Geographers* 56: 39–55.

—— (1987) *The Changing Face of Cities*, Oxford: Basil Blackwell.

Wiebe, R. (1967) *The Search For Order, 1877–1920*, New York: Hill and Wang.

Willensky, E. and White, N. (1988) *AIA Guide to New York City*, third edition, New York: Harcourt Brace Jovanovich.

Williams, B. (1988) *Upscaling Downtown: Stalled gentrification in Washington, DC*, Ithaca and London: Cornell University Press.

Williams, M. (1982) "The new Raj: the gentrifiers and the natives," *New Society* 14 (January): 47–50.

Williams, P. (1976) "The role of institutions in the inner-London housing market: the case of Islington," *Transactions of the Institute of British Geographers* NS 1: 72–82.

—— (1978) "Building societies and the inner city," *Transactions of the Institute of British Geographers* NS 3: 23–34.

—— (1984a) "Economic processes and urban change: an analysis of contemporary patterns of residential restructuring," *Australian Geographical Studies* 22: 39–57.

—— (1984b) "Gentrification in Britain and Europe," in J. J. Palen and B. London (eds.) *Gentrification, Displacement and Neighborhood Revitalization*, Albany: State University of New York Press.

—— (1986) "Class constitution through spatial reconstruction? A re-evaluation of gentrification in Australia, Britain and the United States," in N. Smith and P. Williams (eds.) *Gentrification of the City*, Boston: Allen and Unwin.

Williams, W. (1987) "Rise in values spurs rescue of buildings," *New York Times* April 4.

Wilson, D. (1985) "Institutions and urban revitalization: the case of the J-51 subsidy program in New York City," Ph.D. dissertation, Department of Geography, Rutgers University.

Wines, M. (1988) "Class struggle erupts along Avenue B," *New York Times* August 10.

Winters, C. (1978) "Rejuvenation with character," paper presented to the Association of American Geographers Annual Conference, New Orleans.

Wiseman, C. (1981) "Home sweet Harlem," *New York* March 16.

—— (1983) "The housing squeeze – it's worse than you think," *New York* October 10.

Wolf, E. (1975) *Philadelphia: Portrait of an American City*, Harrisburg, Pa.: Stackpole Books.

Wolfe, J. M., Drover, G. and Skelton, I. (1980) "Inner city real estate activity in Montreal: institutional characteristics of decline," *Canadian Geographer* 24: 349–367.

Wolfe, T. (1988) *Bonfire of the Vanities*, New York: Bantam.

Wright, E. O. (1978) *Class, Crisis and the State*, London: New Left Books.

Wright, G. (1981) *Building the Dream: A Social History of Housing in America*, Cambridge, Mass.: MIT Press.

Wright, H. (1933) "Sinking slums," *Survey Graphic* 22, 8: 417–419.

Wright, P. (1985) *On Living in an Old Country*, London: Verso.

Yates, R. (1992) "Guns and poses," *Time Out* January 8–15, 20–21.

Yeates, M. H. (1965) "Some factors affecting the spatial distribution of Chicago land values, 1910–1960," *Economic Geography* 42, 1: 57–70.

Zukin, S. (1982) *Loft Living: Culture and Capital in Urban Change*, Baltimore: Johns Hopkins University Press.

—— (1987) "Gentrification: culture and capital in the urban core," *American Review of Sociology* 13: 129–147.

Zussman, R. (1984) "The middle levels: engineers and the 'working middle class'," *Politics and Society* 13, 3: 217–237.

INDEX